ADVANCED LEVEL
BIOLOGY
Second edition

Howard Bowen
Joan Good

Series editors
Ted Lister and Janet Renshaw

Stanley Thornes (Publishers) Ltd

Contents

Acknowledgements

The authors and publisher would like to acknowledge the contribution of Ian Simons and Maria Whittington to the first edition.

Thanks are also due to pupils at Trinity School who trialled early versions of this material.

The authors would like to thank the following for their kind permission to reproduce photographs: Geoff Tompkinson / Aspect Picture Lib. (Fig 1.6); M W Tweedie / NHPA (Fig 6.24).

They also thank Mike Bailey, Keith Hirst and Ray Watkin for their kind permission to reproduce, on pages 3 and 4, the diagrams and questions of Activity 55 in *Active Biology: Pupil's Book*, published by Hodder & Stoughton.

First published in 1992 in Great Britain by
Simon and Schuster Education

Second edition published in 1996 by Stanley Thornes (Publishers) Ltd
Ellenborough House
Wellington Street
Cheltenham
Gloucestershire GL50 1YW

97 98 99 00 / 10 9 8 7 6 5 4 3 2

A catalogue record for this book is available from the British Library
ISBN 0-7487-2333-1

Typeset by Techset Composition Ltd
Printed and bound at Redwood Books, Trowbridge, Wiltshire

Introduction

This book is designed to help you get ready for a post-16 course in biology: Advanced Level, Advanced Supplementary Level, BTEC, GNVQ, Scottish Higher, etc. It doesn't matter which course you will be following because this book stresses the principles of biology, which are the same for any course.

You can use this book before you begin your advanced course or during the first part of the course. You will also find it useful if you are aiming for the higher level GCSE papers. It will also be useful for reference during your advanced course.* *Access to Advanced Level: Biology* has been designed so that you can work through it on your own, so the answers to all the exercises are at the end of the book. However, cheating won't help your understanding!

How to use this book

Teaching yourself how to do something needs confidence, which often needs developing. We suggest that you work through each section slowly and don't move on to the next section until you have correctly answered the exercises. If you are getting most of them right you are doing well. But what if you aren't? One of the skills you will have to develop for advanced study is independent learning. There is a variety of approaches to explaining concepts and ours may not always be the best for you. At advanced level (and beyond), using other resources, such as standard text books, and reading on your own initiative, is going to give you a valuable skill and that essential ingredient, **confidence** in your ability to learn by yourself.

Good luck and enjoy your biology!

The authors and editors

* There is a glossary of important terms at the back of the book for you to refer to.

Chapter 1 ▶ Energy in the living world

STARTING POINTS

- Do you understand the following terms: **photosynthesis**, **respiration**, **metabolism**? Write one or two sentences (no more) to explain what you understand by each term. Then check them against the glossary at the back of the book.

Energy from the Sun

All living organisms need a constant source of energy. All of this energy ultimately comes from the Sun.

Green plants have **chlorophyll**, a pigment that can absorb sunlight energy and store it in a chemical form (food) made by the process of **photosynthesis**. Plants build up and store food in this way for themselves and the rest of the living world, so they are called **producers**. Organisms that do not have such a pigment cannot tap the Sun's energy and must obtain energy from the food that they eat.

Animals obtain their energy by eating food that has been made by plants. They are called **consumers**. Transfer of energy in the form of food from plants to a variety of animals forms the basis of **food chains**. In plants and animals, energy to carry on cell processes comes from food, whether the food is made within the organism by photosynthesis or taken in from the environment (eaten).

An oak tree is a producer organism, but many cells in its roots, its trunk and its branches – all those cells that do not contain chlorophyll – function like the cells in a consumer. Even its green cells are consumers in the dark. They depend upon food made by photosynthesis until they are in the sunlight again. The release of energy from food occurs in all cells at all times, as long as they are alive, by the process of **respiration**.

Exercise

> 1 Divide the following into producers and consumers: dog, thistle, blackbird, seaweed, coral, algae, mushroom.

The theme of this chapter is the energy relationships of the living world that exist through the processes of photosynthesis and respiration. Chapter 2 looks at the spread of this energy through food chains.

Photosynthesis

All living cells need energy. Except for a few bacteria that obtain energy from chemical reactions, all living cells use energy that originally came from the Sun. Only the green parts of plants can trap this solar energy whilst they carry out photosynthesis.

The discovery of photosynthesis

Read the passage and then answer the questions in the following exercise.

In 1772 Joseph Priestley placed a shoot of mint in a container of water and inverted a glass jar over it so that air could not enter. To his surprise the shoot remained alive for several months. In another experiment he noted that a burning candle was quickly extinguished when covered with a jar. Then Priestley placed a shoot of mint under the jar, and in a few days the candle, when lighted, burned again for a short time.

The 'restored' air was, as he said, "not at all inconvenient to a mouse which I put into it". Other investigators, however, were unsuccessful when they attempted to repeat Priestley's experiments.

The reason for this failure became clear in 1779, when Jan Ingen-Housz found that plants behave in the way Priestley described only when they are exposed to sunlight.

In 1782 Senebier discovered that plants absorb carbon dioxide when in sunlight. In 1804 de Saussure showed that the increase in plant weight after exposure to sunlight is greater than the weight of the carbon dioxide taken in.

By 1845, Mayer was able to recognise that the essential steps in photosynthesis are the absorption of light energy and the transformation of this light energy into chemical energy, which is then stored in chemical compounds manufactured by the plant. The process can be represented by the following equation:

$$\text{carbon dioxide } + \text{ water } \xrightarrow[\text{chlorophyll}]{\text{sunlight}} \text{ glucose } + \text{ oxygen}$$

$$6\,CO_2 \quad\quad + \ 6\,H_2O \quad \rightarrow \quad\quad C_6H_{12}O_6 \ + \ 6\,O_2$$

Exercise

2 a What is the name of the gas that 'restored' the air in Priestley's experiment?
  b Why did the mint plant only restore the air when it was in sunlight?
  c What other substance is taken in by plants during photosynthesis to explain the further increase in plant weight observed by de Saussure?

Oxygen production in photosynthesis

In the 1880s a German biologist called Englemann carried out one of the first 'bioassay' experiments in science. A bioassay involves using a living organism to detect or measure some factor present in minute amounts. He used a freshwater bacterium which can swim and move to regions of high oxygen concentration. In a number of ingenious experiments on photosynthesis, Englemann placed a culture of the bacteria in test tubes with a freshwater alga (*Spirogyra*) under different light conditions. Fig 1.1 shows what he saw when he kept the *Spirogyra* and bacteria (a) in bright light, (b) in the dark, (c) in red light, (d) in green light, and (e) in blue light.

Exercise

3 a What do you notice about the distribution of the bacteria
    i in the light
    ii in the dark?
  b Which part of the *Spirogyra* cell is responsible for producing oxygen? Explain how you worked this out.
  c The oxygen is being produced by photosynthesis. Which colours of light are shown in the experiment to be used for photosynthesis?
  d Why do you think that green light does not bring about photosynthesis?

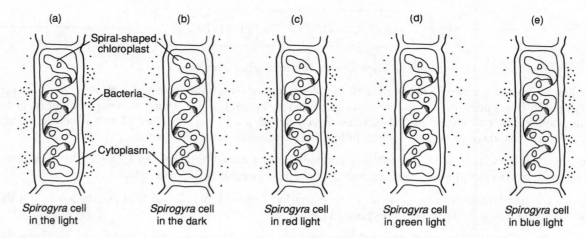

1.1   The alga *Spirogyra* and the positions of swimming bacteria in five different light conditions.

The mechanism of photosynthesis

To demonstrate conclusively that photosynthesis occurs only in the **chloroplasts**, we need to separate the chloroplasts from the rest of the cell. Hill managed to do this in 1937 and in 1954 Arnon proved that chloroplasts, separated from all other parts of the cell, can carry out the entire process of photosynthesis.

Photosynthesis is divided into two phases:

- **Light reaction** – chlorophyll absorbs light energy which is converted to chemical energy. Water molecules are split, oxygen is given off, and the hydrogen is used to react with carbon dioxide in the second phase. This phase needs light but is not affected by temperature.

- **Dark reaction** (this does not require light) – carbon dioxide is converted into sugars by a series of reactions. This phase, because it is made up of chemical reactions, is affected by temperature; the light reaction is not. An increase in temperature of 10 °C will approximately double the rate of a chemical reaction.

Exercise

4 a A geranium plant is placed in full sunlight. If the temperature around the plant is increased from 10 °C to 20 °C what effect will this have on the rate of the
  i light reaction
  ii dark reaction?
 b Another plant in full sunlight has the temperature around it reduced from 25 °C to 15 °C. What effect will this have on the rate of the
  i light reaction
  ii dark reaction?

The rate of photosynthesis

This depends on several factors but the four main environmental ones are the following, in order of priority:

a the light intensity falling on the plant
b the temperature, which affects the dark reaction. The optimum temperature is about 30 °C
c the amount of carbon dioxide in the atmosphere (usually 0.04%)
d the amount of water available to the plant, but this only becomes important during severe drought conditions.

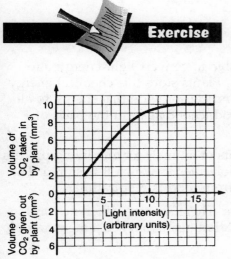

1.2 Graph showing the effect of different light intensities on the carbon dioxide exchange of a plant.

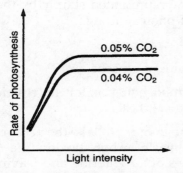

1.3 Graph showing the effect of different light intensities on the rate of photosynthesis at two different concentrations of carbon dioxide.

5 Fig 1.2 shows the effect of increasing the light intensity on the uptake of carbon dioxide by a green plant.

a How many units of light was the plant receiving when it took in 6 mm^3 of carbon dioxide?

b What increase in carbon dioxide uptake is caused by increasing the light intensity from
  i 3 units to 6 units
  ii 10 units to 13 units?

c Use a ruler to help you work out the light intensity at which carbon dioxide would be neither taken in nor given out by the plant.

The limiting factors of photosynthesis

Any one of the four factors labelled a–d above may limit the rate at which a plant can photosynthesise, provided all other factors are in excess. In dull weather, light is often the limiting factor so the rate of photosynthesis can be increased by increasing the light available. In a sunny climate, lack of carbon dioxide may limit the rate.

Fig 1.3 shows the rate of photosynthesis for a series of different light intensities. The rate rises sharply and then flattens out at a maximum.

An increase in light intensity beyond this point does not increase the rate further. Some other factor, probably carbon dioxide or temperature, is acting as a limiting factor at this level of light intensity. Repeating the experiment with one factor altered (for example an increased level of carbon dioxide) may cause a further increase in the rate of photosynthesis. If it does, as in Fig 1.3, then the altered condition was previously the limiting factor.

6 Look at the graphs in Fig 1.4 which show the rate of photosynthesis under different conditions. At each plateau on the graphs, a factor is limiting photosynthesis.

Name the most likely limiting factor at points A, B, C and D on the graph.

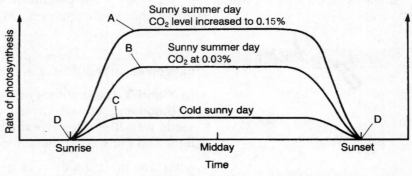

1.4 Graph showing the rate of photosynthesis at different times of the day under three different conditions.

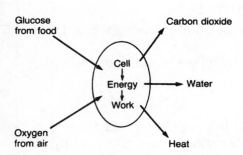

1.5 Exchanges carried out by a cell during respiration.

1.6 A spirometer.

Respiration

We have just seen that plants are able to convert light energy into chemical energy by photosynthesis. Plants store this chemical energy as foods such as starch or sugars and can provide this energy to all other forms of life. They are the producers that form the basis of all **food chains**.

This energy cannot be used by plants or animals for any living process until it is released from the food molecules in which it is stored. The process which releases the energy from food is **respiration**.

All living organisms have to respire. There are two different ways in which they do this – **aerobic respiration** (using oxygen) and **anaerobic respiration** (without oxygen).

Aerobic respiration

This occurs in most organisms and requires a constant supply of oxygen. The raw materials enter the cell, respiration occurs and waste products are removed (Fig 1.5). It can be represented simply by the chemical equation that is the reverse of photosynthesis.

$$glucose + oxygen \rightarrow carbon\ dioxide + water$$

This has an energy output of 2800 kJ/mol of glucose.

See if you can work out the balanced symbol equation for this reaction. **Hint:** look in the section above on photosynthesis.

To respire aerobically, we need to **breathe** in order to take the necessary oxygen into our bodies and release the waste carbon dioxide (see Chapter 4). We can measure the volumes of air entering and leaving our lungs in each breath using a piece of apparatus called a spirometer – see Fig 1.6. A pen moves up and down on a revolving drum making a trace as the person breathes in and out. Fig 1.7 represents such a trace.

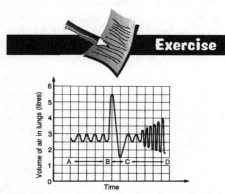

Exercise

1.7 Spirometer trace of a subject starting at rest (A–B).

Example

7 a In which direction does the pen move when air is being
   i inhaled
   ii exhaled?
 b What volume of air is inhaled and exhaled at each breath during period A–B when the person is at rest?
 c Describe in detail the breathing pattern of the subject during period B–C on the graph.
 d Work out the maximum amount of air that this person can exhale in one breath as indicated by the graph.
 e What is the person likely to be doing to cause the graph to change during period C–D on the graph?

An experiment was set up, as shown in Fig 1.8, to find the volume of oxygen absorbed during aerobic respiration by germinating pea seeds.

In X and Y, the volumes of the test tubes were the same. The cross-sectional area of each capillary tube was 3 mm². The weight of the seeds was 10 g. The experiment was carried out at 25 °C in uniform light and constant atmospheric pressure and temperature.

After one hour the drop of coloured liquid in the capillary tube X had not moved. In Y the drop in the capillary tube had moved 50 mm from its position at the start of the experiment in the direction of the test tube.

a What must have caused the movement of the drop in capillary tube Y?

Answer: The seeds respired and exchanged oxygen for carbon dioxide. The carbon dioxide was absorbed by the potassium hydroxide solution so that the coloured liquid in the capillary tube was drawn into tube Y.

b Why do you think perforated platforms were used?

Answer: To allow gas movement throughout each tube.

c Why do you think 5 cm³ of glass beads were placed in tube X?

Answer: These beads occupied the same amount of space in tube X as the seeds in tube Y, so that both tubes contained the same amount of air at the beginning of the experiment.

d Work out

i the volume of oxygen (in mm³) which the seeds used in one hour.

Answer: The volume of oxygen absorbed is equal to the volume of the capillary tube along which the coloured liquid moved.

This is $50 \text{ mm} \times 1 \text{ mm}^2 = 50 \text{ mm}^3$.

ii the volume of oxygen absorbed per gram of seed.

Answer: 50 mm³ of oxygen was used by 10 g of seed, so $50/10 = 5 \text{ mm}^3$ of oxygen was used per gram of seed.

e What would you expect to happen to the rate of oxygen uptake by the seeds if the experiment were carried out

i at 15 °C in uniform light?

Answer: Respiration would be slower at this lower temperature so the liquid in capillary tube Y would move more slowly. (Since respiration is a series of chemical reactions, the reduction in temperature of 10 °C would have approximately halved the respiration rate.)

ii at 25 °C in the dark?

Answer: There is no change in temperature and respiration would go on at the same rate in the dark as in the light. (Respiration is not affected by light/dark conditions.)

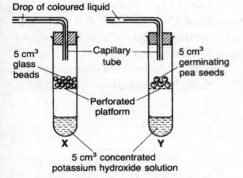

Drop of coloured liquid

5 cm³ glass beads

Capillary tube

5 cm³ germinating pea seeds

Perforated platform

X   Y

5 cm³ concentrated potassium hydroxide solution

1.8 Apparatus used to find the volume of oxygen absorbed during aerobic respiration by germinating pea seeds.

Exercise

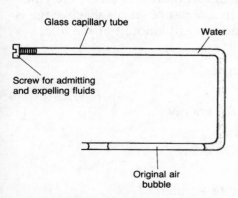

Glass capillary tube

Water

Screw for admitting and expelling fluids

Original air bubble

1.9 Apparatus used to investigate the composition of a bubble of exhaled air.

8 The apparatus in Fig 1.9 was used to investigate the composition of a bubble of air (exhaled) after exercise. With the bubble always at constant temperature and pressure, the following results were obtained:

Length of gas bubble in contact with water	= 100 mm
Length of bubble after contact with potassium hydroxide (which removes carbon dioxide)	= 94.5 mm
Length of bubble after contact with potassium pyrogallol (which removes oxygen)	= 79 mm

a Work out the percentage of carbon dioxide present in the original bubble.

b Work out the percentage of oxygen present in the original bubble.

c Use a text book (in a library search) to find out how these percentages compare with the amounts of carbon dioxide and oxygen in the atmosphere.

Chemical reactions in the cell

The living cell has many chemical reactions going on in the cytoplasm at any one time. These need to occur quickly. In the cell chemical reactions are speeded up by using **enzymes**. These are biological catalysts made of protein. They reduce the **activation energies** of the reactions. Enzymes are very specific, usually catalysing one reaction of one compound called the **substrate**. The enzyme molecule has a particular shape built into its surface into which both the substrate and the molecule with which it is reacting can fit. The close proximity of the two reacting molecules induces them to react. The products are then released by the enzyme.

Because the catalytic activity of enzymes is so dependent on their shape, enzymes are very sensitive to changes in pH and temperature which can cause them to **denature** (change their shape). An enzyme has an optimum pH and an optimum temperature at which it works best. Below the optimum temperature, the reaction is slow as the substrate molecules are not moving very fast. Above this temperature, there is distortion of the shape of the enzyme molecule and it loses its catalytic effect – see Fig 1.10.

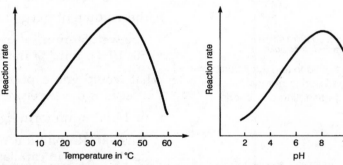

1.10 Graphs of reaction rate against temperature and pH for an enzyme-catalysed reaction.

 **Exercise**

9 a What is i the optimum pH and ii the optimum temperature, for the enzyme used in Fig 1.10?
  b Explain why the rate of the reaction decreases so rapidly at temperatures higher than 50 °C.

The storage of energy in the cell

Inside the cytoplasm and the many **mitochondria** of a cell, enzymes break down the glucose molecules in a series of stages. Each stage is controlled by a different enzyme and a small amount of the total energy of the molecule is released.

Glucose

enzyme 1 ↓

Substance B + energy

enzyme 2 ↓

Substance C + energy

several more stages ↓

Carbon dioxide + water

(Total energy output = 2800 kJ/mol of glucose)

In the mitochondria of each cell this energy is used to make molecules of the energy-rich compound called **adenosine triphosphate (ATP)** as follows:

$$\text{energy from respiration}$$

adenosine diphosphate (ADP) + phospate → adenosine triphosphate (ATP)

$$A-\textcircled{P}-\textcircled{P} \qquad +\textcircled{P} \qquad \rightarrow \qquad A-\textcircled{P}-\textcircled{P}\sim\textcircled{P}$$

(where $\textcircled{P}$ = phosphate group)

ATP is the intermediate energy compound in the cell. If the energy in the glucose molecules were released suddenly, the cell would be destroyed by the excess heat. When released slowly, via ATP, most of the energy can be harnessed for the work of the cell.

The chemical bond $\textcircled{P}\sim\textcircled{P}$ is 'energy-rich' and when broken down provides energy for the work of the cell:

$$\text{ATP} \rightarrow \text{ADP} + \textcircled{P} + \text{energy for work}$$

The amount of energy released by an organism in a given time is known as its **metabolic rate**. Whenever energy is transferred from one compound to another or used in work, some is lost as heat energy. This is usually considered as lost or wasted energy but warm-blooded animals, such as mammals, can control their heat loss and use it to maintain a constant body temperature.

ATP is the first of several different levels of energy storage in the organism. A useful analogy of these energy stores is the use that we make of our 'money stores' when we need to buy something.

Money store 1: A small purchase in a shop can be made simply by using the small change in our pocket or purse. This cash is analogous to the ATP energy available to a cell. It is present in small amounts, but is readily available.

Money store 2: For a more expensive article we are going to have to break into the ten pound note in our wallet. The change from this note will help replenish our money store 1. In the cell, activities requiring more energy will require energy from the respiration of glucose which can be stored as ATP as explained above.

Money store 3: Sometimes the contents of our wallet are not enough. We now need to get extra money from our current account in the bank or building society. In the organism this is similar to raiding the energy stores of larger carbohydrate molecules such as starch in plants, or glycogen stored in the liver of animals. Both these molecules can be converted into glucose which is then respired.

Money store 4: A really expensive article such as a car means that we now have to raid long term savings and deposit accounts. In an animal this is similar to now using the fat deposits around the body to obtain the required energy. The fat has been stored for a long time and is only used when the animal runs out of the carbohydrate energy store. This is, of course, the principle of going on a diet in order to lose weight.

Money store 5: Our financial state is becoming drastic and we are now having to sell off family treasures in order to obtain the required money. In the organism, after carbohydrate and fat stores have run out, there is only one more possible energy store. This is the protein from which the structure of the body is built. In these dire circumstances – a starvation diet – the muscles of the animal will waste away in order to provide the last bit of energy needed to keep the animal alive.

Surface area to volume ratio

The surface area to volume ratio of an animal is important in determining how quickly it will lose heat from its body. Animals with large surface area to volume ratios will lose their body heat faster than animals with smaller surface area to volume ratios.

Area of face = 1 × 1   = 1 cm²
Total area   = 6 × 1   = 6 cm²
Volume   = 1 × 1 × 1 = 1 cm³

1.11   Areas and volume of a cube.

Exercise

10 The table below, when completed, shows the relationship between surface area and volume in a series of cubes of different sizes (Fig 1.11).

Length of side of cube (cm)	Total surface area (cm²)	Volume (cm³)	Surface area/ volume ratio
1	6	1	6 : 1
2	?	?	?
3	?	?	?

a Copy out the table, inserting the correct figures in the places marked '?'.
b What happens to the surface area to volume ratio as the cubes increase in size?
c The African elephant, which lives in a hot climate
   • has almost hairless skin
   • can increase its surface area by approximately one-sixth by raising its large ears.
   How do you think each of these facts helps the elephant to control its body temperature?
d An adder is said to be 'cold-blooded' whilst mammals are 'warm-blooded'. Use a reference book to help you answer the following questions.
   i   Explain what is meant by the term 'warm-blooded'.
   ii Explain two ways in which being 'warm-blooded' has advantages over being 'cold-blooded'.
   iii Apart from mammals, name one other animal group (phylum) that is 'warm-blooded'.

The size of cells

The surface area to volume ratio is also important in determining the size of cells. Figure 1.5 showed some of the major materials that a cell has to exchange with its environment in order to live.

A very large cell would have a small surface area compared with its volume. The volume of a cell determines the amount of materials that the cell would need to absorb through its surface. Its comparatively small surface area would mean it would not be able to absorb these materials in sufficient quantity.

A very small cell with a large surface area to volume ratio would be able to absorb these materials in the quantities required, but the small cell would be limited in the number of functions that it could carry out. There would not be enough room in the cell to house the many organelles required for many different functions.

Consequently, the cells of higher organisms (plants and animals) are all about the same optimum size. Their cells are large enough to contain different organelles such as **mitochondria** and **chloroplasts** and have a surface area to volume ratio that is sufficient for them to obtain all their required materials through their cell membrane.

Simpler organisms, such as bacteria and blue–green algae are made of single, smaller cells so that their large surface area to volume ratio easily enables them to obtain all materials that they need. Their small size, however, means that there is not enough room for their cells to contain organelles and this has restricted them to a primitive lifestyle.

Anaerobic respiration

The breakdown of glucose *without* oxygen occurs in some fungi and bacteria, and also in muscles during vigorous exercise. It is not as efficient as aerobic respiration in terms of energy production and ATP yield.

Anaerobic respiration in yeast (fermentation)

This process is important in the baking and brewing industries. Single-celled yeasts grow and divide very rapidly and obtain their energy by breaking down sugars as shown below:

glucose $\rightarrow$ ethanol + carbon dioxide

$C_6H_{12}O_6$ $\rightarrow$ $2\,C_2H_5OH$ + $2\,CO_2$ + energy

Energy output = 200 kJ/mol of glucose

Ethanol and carbon dioxide are the waste products. Carbon dioxide is used to make dough rise and the alcohol produced by brewer's yeast is used in beer- and wine-making.

'Gasohol'

Some South American countries have no major oil reserves and cannot afford to import oil. These tropical countries are able to grow a lot of sugar cane and excess sugar from the crop can be fermented to alcohol. Alcohol can be used instead of petrol in cars. Blended with some petrol it is sold as the fuel 'gasohol' or 'alcool'.

In Brazil many cars now run on alcohol fuel made in this way and Brazil has plans eventually to replace all petrol by alcohol. The low-revving engine of the Volkswagen Beetle runs well on this fuel.

Anaerobic respiration in muscle

During vigorous exercise the oxygen supply to the cells becomes insufficient despite faster breathing. Anaerobic respiration occurs at the same time as aerobic and supplies extra energy for muscle contraction. Lactic acid accumulates and an **oxygen debt** is built up.

glucose $\rightarrow$ lactic acid + energy

The lactic acid enters the bloodstream and passes to the liver where it can be broken down to other substances. The reactions that get rid of the lactic acid use up oxygen, and this makes up the 'oxygen debt' which now has to be repaid.

Exercise

11 a When do our muscles respire anaerobically?
   b Explain why this cannot go on indefinitely.
   c Why do the following animals need to respire anaerobically for long periods of time?
      i tapeworms   ii seals   iii mud-burrowing worms

Gas exchange by plants

The overall exchange of gas between plants and the atmosphere depends on the light conditions.

During daylight, plants carry out photosynthesis and respiration. The rate of photosynthesis is much greater than the rate of respiration and supplies all the oxygen for energy release. It also uses up the waste carbon dioxide from respiration. During daylight, gas exchange with the environment involves the overall uptake of carbon dioxide and release of oxygen.

During darkness plants carry out respiration only. They then take up oxygen and release carbon dioxide.

All plants and animals respire at all times. Green plants photosynthesise during daylight hours only.

The compensation point

At two points every 24 hours, gas exchange through the tiny holes in the leaf surface, called **stomata**, seems to stop. These are the **compensation points** and they are the times at which the rates of photosynthesis and respiration are equal (see Fig 1.12).

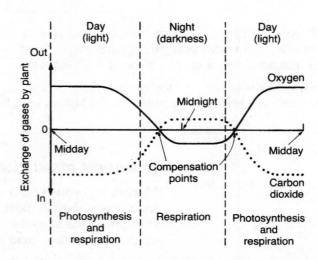

1.12   Exchange of oxygen and carbon dioxide by a plant over a 24-hour period.

Exercise

12 Fig 1.13 shows the effect of increasing light intensity during the day on the gas exchange of two woodland plants – an oak tree which has a leaf canopy in full sunlight and a bluebell which lives in the shade of the oak tree.

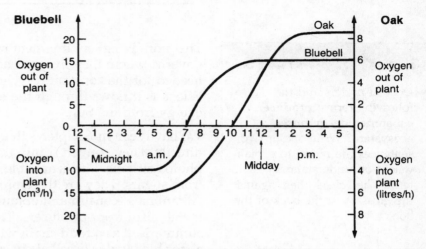

1.13   Effect of changing light intensities during the day on the rate of oxygen exchange by an oak tree and a bluebell.

   a At which time of day is the compensation point achieved by the
     i oak tree   ii bluebell?
   b Why do you think it is an advantage for the bluebell to reach its compensation point at a lower light intensity than the oak tree?
   c Why do you think the graph for the oak tree reaches a higher plateau than the graph for the bluebell?
   d How much oxygen does the bluebell require for respiration between 2 a.m. and 3 a.m?
   e How much oxygen does the bluebell produce by photo-synthesis between 2 p.m. and 3 p.m?

13 At Christmas a garden centre sold 'bottle-gardens'. These were sealed, transparent glass containers containing some moist compost rooted into which was a variety of plants. They were advertised as being an excellent present for the lazy gardener as, apart from putting them in sunlight, they required no other maintenance. Such a gardener received the bottle-garden present and put it in plenty of light in her warm conservatory. The plants survived in the bottle but never grew. After several months the gardener began to feel sorry for the plants and removed the lid that sealed the bottle. The plants now began to grow and were soon too big for the bottle.
Explain why the plants

   a did not grow in the sealed bottle,
   b did grow after the seal was broken.

Chapter 2

Food chains

STARTING POINTS

● Do you understand the following terms: **producer, consumer, decomposer, ecosystem**? Write one or two sentences (no more) to explain what you understand by each term. Then check them against the glossary at the back of the book.

The 'non-living' environment supports living organisms. It therefore governs where they live and how many there are of them. Energy is needed for the constant recycling of materials within an **ecosystem**. It is lost as it flows through the ecosystem and must be replaced by energy from the Sun.

We have seen in Chapter 1 that only green plants can use solar energy directly. They convert it into the chemical energy of sugars by photosynthesis. Plants are able to build up proteins, fats and vitamins from sugars. They are therefore called the **producers** of the ecosystem. All animals (**consumers**) obtain their energy for growth and metabolism from producers. These feeding relationships can be summarised as a **food chain** where each stage of the chain is known as a **trophic level**. Microbial consumers, which aid the decay of dead organisms at all levels, are called **decomposers**. The following diagram shows a typical food chain:

3rd trophic level	Lion	Carnivore: second consumer
	↑	
2nd trophic level	Antelope	Herbivore: primary consumer
	↑	
1st trophic level	Savannah grass	Producer
	↑	
	Solar energy	

Only a small amount of the total energy that reaches the plant as light is incorporated into plant tissues. As energy is passed along the food chain there is a large loss between each level.

It is rare to find a simple food chain in an ecosystem. Usually there are several organisms at each level which may obtain food from any one of the lower levels. These complex feeding interrelationships are called **food webs**.

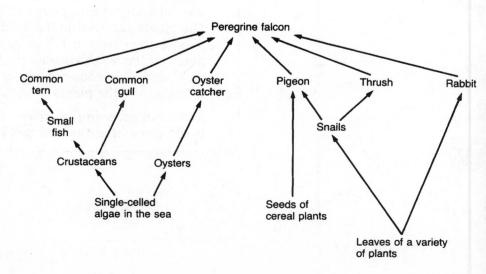

2.1   Food web for the peregrine falcon.

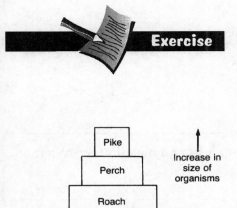

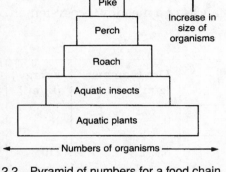

2.2 Pyramid of numbers for a food chain in a pond.

1 The food web in Fig 2.1 shows some of the feeding relationships of the peregrine falcon. It is a bird of prey which lives and nests on cliffs by the sea and feeds almost entirely on other birds

  a What is the ultimate source of energy for all the organisms in this food web?

  b Name one organism from this food web which is a producer.

  c What would happen to the common tern population if the numbers of common gulls were reduced?

  d What would happen to the thrush population if the numbers of pigeons were increased?

  e How is energy lost from each stage of this food web?

Pyramids of numbers

There is generally an increase in size (mass) of organism from primary consumer to the final carnivores in a food chain, but a decrease in numbers (see Fig 2.2). As there is a decrease in available energy at each successive link in the food chain there must also be a decrease in the amount of living material that it can sustain.

2 Fig 2.3 shows different patterns of pyramids of numbers for different food chains. The producers are shown as the bottom of each pyramid. The pyramids are not drawn to scale.

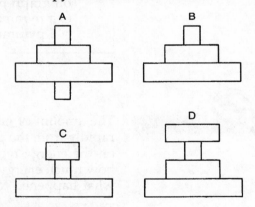

2.3 Different pyramids of numbers.

Match each of the following food chains with one of the pyramids of numbers A–D.

i   rose bush → greenfly → ladybirds → great tit

ii African savannah grass → wildebeest → lion

iii leaf litter → earthworms → hedgehog → fleas

iv waterhole vegetation → water buffalo → mosquitoes

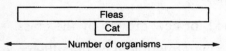

2.4 Pyramid of numbers for the cat → flea food chain.

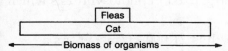

2.5 Pyramid of biomass for the cat → flea food chain.

Pyramids of biomass and energy

Some of the pyramids of numbers in exercise 2 are only pyramid shaped because we are counting just the *numbers* of organisms. Sometimes the pattern of numbers is reversed, for example, there may be around one hundred fleas feeding on the blood of one cat. A pyramid of numbers for this situation is shown in Fig 2.4.

However, if we measure the total mass of these two organisms we get a true pyramid pattern – a pyramid of **biomass** (the living weight of the organisms) as shown in Fig 2.5.

3 The following table shows an analysis of a meadow ecosystem in Britain.

Trophic level	Organisms	Dry biomass (g/m²)	Energy (kJ/m²)
Top carnivore	Kestrel	0.5	16
Secondary consumers	Sparrows, tits	8.4	210
Primary consumers	Insects, molluscs	42.5	640
Producers	Meadow vegetation	765.0	13 200

a Why it is better to use dry biomass as a measure rather than fresh biomass?

b For this meadow ecosystem draw accurate scaled diagrams on graph paper to show
   i a pyramid of biomass
   ii a pyramid of energy.

The amount of energy in each trophic level of a food chain decreases rapidly from the producers to the top carnivores. Below are some data on the energy budget for an agricultural field ecosystem. This shows how much energy each square metre of field received in one year, and what happened to the energy used by the grass in the field.

Total energy received from sunlight   1 000 000 kJ/m²/yr.
Total energy gained by photosynthesis   50 000 kJ/m²/yr.
Total energy lost through respiration   29 000 kJ/m²/yr.

a Calculate the percentage of the sunlight energy received which is fixed by the grass in photosynthesis.

   Answer: (50 000/1 000 000) × 100 = 5%.

b Calculate how much energy is used for new growth by the grass.

   Answer: The energy available for new growth will be the difference between the energy provided by photosynthesis and the energy used by respiration. This is 50 000 − 29 000 = 21 000 kJ/m²/yr.

c Calculate the percentage of the energy fixed in photosynthesis which is used for new growth.

   Answer: The energy for new growth is 21 000 kJ/m²/yr. As a percentage of energy fixed by photosynthesis this is (21 000/50 000) × 100 = 42%.

d Suggest what happens to the energy from sunlight which is received by the grass but is not used in photosynthesis.

Answer: Much of the energy in sunlight is not light energy which can be used for photosynthesis. For example, the heat energy in the sunlight is not directly used for photosynthesis.

This pattern of only a small amount of the energy taken in by an organism being used for growth is repeated at each trophic level.

Exercise

4 Here is some information about the energy available to a bullock feeding on the grass in the field referred to above.

Energy in grass eaten by the bullock   3050 kJ/m^2/yr.

Energy released in respiration   1020 kJ/m^2/yr.

Energy lost in faeces and urine   1900 kJ/m^2/yr.

a Work out the percentage of the available energy in the grass which is eaten by the bullock.

b Suggest why the bullock is only able to make use of such a low proportion of the available energy in the grass.

c What percentage of the energy actually taken in by the bullock is used for new growth?

d In terms of food efficiency, explain why the raising of bullocks in a field for meat produces so little food compared with using the field to grow vegetables.

e Why does our diet contain more meat from herbivores than from carnivores?

f In some tropical regions, native herbivores such as zebra, wildebeest and antelope yield more energy per unit area than introduced species of cattle. Suggest some reasons for this.

Variation in numbers within a population

Populations of organisms vary in response to changing conditions within an ecosystem, but over a long period of time the average size of each population remains the same.

If the predators become too numerous they will reduce the numbers of herbivores. Then the predator numbers will decline because of food shortage. Then, with fewer predators the herbivore numbers will again begin to rise. Over a period of time there is a general balance in numbers of predator and prey. It is not an absolutely steady state and is often called a 'state of dynamic equilibrium' in the ecosystem.

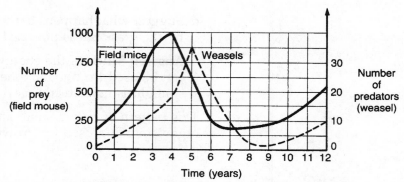

2.6   Graph showing population changes for weasels and field mice over a 12-year period.

5 Fig 2.6 shows the results from a study of the populations of two organisms, the weasel and its prey, the field mouse, in an enclosed farm area over a period of 12 years.

a How many years did the population of field mice take to reach its maximum size?

b How many years did the population of weasels take to go from maximum to minimum numbers?

c How many times greater was the total number of weasels present in year 3 than the number present at the start of the study?

d Try to account for the decrease in field mice and the increase in weasels during the period between year 4 and year 5.

Changes in populations with time

Each organism lives in its own 'niche' in which it finds the range of environmental factors to which it is adapted. However environmental factors (such as the humus and water content of soil) do not necessarily remain constant. Even the activities of the organism itself can alter the surrounding environmental factors so that the niche is no longer suitable for it. Other organisms can then move in to take its place. This replacement of populations is known as **succession** and in its final stage produces the **climax community** which is the stable community of organisms in equilibrium with that particular environment. The climax community is determined largely by the climate so that over most of Britain the climax vegetation would be deciduous woodland such as oak and beech (Fig 2.7).

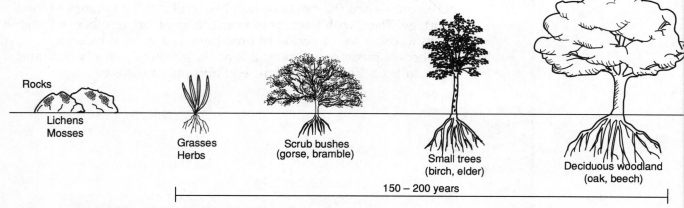

2.7   Typical vegetation succession in the southern U.K.

Sand dunes – an example of succession

6 Read the passage and then answer the questions which follow.

It is difficult to live just above high tide on a sandy beach. Although it is beyond the range of most tides, it receives salt spray in the wind and the sand is loose, unstable and holds very little rain water.

One plant, sand couch grass, can withstand these conditions and also the occasional immersion in sea-water. It is the **pioneer plant** on this part of the beach and soon wind-blown sand gets trapped around the plants and a sand dune begins to form.

Gradually the dune builds up until it is high enough to be out of reach of the sea. Marram grass now moves in and ousts the sand couch grass. It cannot tolerate any sea-water immersion but it has a stronger and deeper root system to anchor it into the sand and to obtain what water is available. It thrives on 'mobile' dunes where fresh sand is periodically blown over it. It grows through the wind-blown sand making the dune more stable and this is its eventual downfall.

Behind the mobile dunes the wind speed drops, there is much less wind-blown sand, and the sand, now made stable by the marram grass, provides anchorage for opportunist plants to colonise the dune. Small animals can live amongst the vegetation. Animal faeces and the remains of the dead animals and plants begin to build up the humus content of the dune. The sandy soil now retains water and nutrients better and these darken the soil to that of a **grey dune**. Over hundreds of years, mosses, lichens, flowering plants, herbs and shrubs colonise the dune system in succession.

a What features allow the sand couch grass to be the pioneer plant just above the high tide level of the beach?
b List the prevalent environmental conditions in the mobile dunes.
c Both storms and human erosion can cause 'blow-outs' in the dune system in which sand can be blown out of the dune by the wind. What treatment could be carried out to re-establish the blow-out back to the normal dune system?
d Give the reasons for the marram grass being ousted by other plants from the mature dunes.

Recycling of materials

A characteristic shared by living organisms is that they all eventually die. The death and decay of living things is essential in nature as it allows for the recycling of certain elements.

Plants take carbon dioxide into their leaves and fix the carbon into carbohydrate. They absorb nitrates into their roots and combine this with carbohydrate to make protein. These materials pass down food chains and on the death and decomposition of the organisms they are returned to the 'natural pool' for recycling. Figs 2.8 and 2.9 show some of the processes involved in the two major biogeochemical cycles – the nitrogen cycle and the carbon cycle.

Answer the exercise on the carbon cycle after reading through the example on the nitrogen cycle.

a Name the four major elements that make up plant protein and describe in what form the plant obtains these elements.

Answer: Carbon (carbon dioxide), hydrogen (water), oxygen (water and carbon dioxide), nitrogen (nitrates in soil except for legumes which can fix nitrogen from the air).

b Explain why carnivorous plants are found in areas where the soil lacks available nitrates.

Answer: They obtain their nitrogen from animals.

c Peat is made up of partially decomposed bog plant remains. Suggest why this peat accumulates on cold moorland such as the hillsides of Wales and Ireland.

Answer: The low temperature inhibits decomposition by bacterial action.

d Since medieval times farmers have included the growth of legumes such as clover in their crop rotations. Suggest a reason why modern 'organic' farmers have to do this.

Answer: Legumes fix nitrogen and this reduces the need for artificial fertilisers which organic farmers do not use.

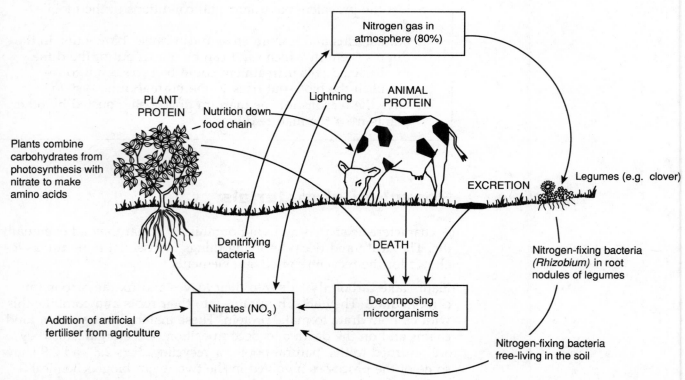

2.8   The nitrogen cycle.

Exercise

7 Solid lines in Fig 2.9 show that the carbon pathway occurs constantly and rapidly. Pathways shown by dotted lines can lock up the carbon for long periods of time (millions of years).

a Which two types of organisms are the main constituents of the decomposer group?
b Explain the process by which carbon has become locked up as limestone. Which organisms in the sea were the last link in the production of limestone?
c What industrial building material is produced in large amounts from the burning of limestone?
d Use the pathways from the carbon cycle to explain how human activity is causing an increase in the amount of carbon dioxide in the atmosphere.

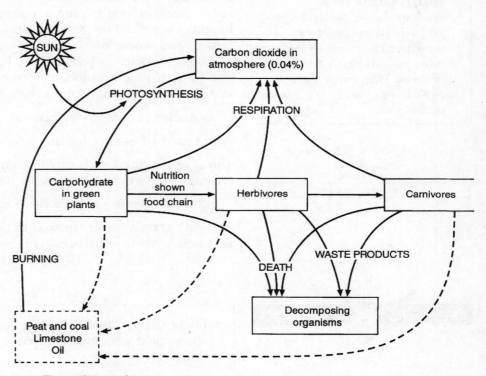

2.9   The carbon cycle.

Chapter 3

Effect of humans on the environment

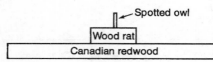

3.1 A pyramid of energy.

Food chains and human population growth

The largest organisms in the world are the Canadian redwood (Giant Sequoia) trees of northern USA and Canada. The leaves of the redwood are inedible so most of its photosynthetic food production is not passed along to animal consumers. The redwood tree grows to heights of over 80 metres and attains masses of around 2500 tonnes. Leaves and whole trees eventually die and fall to the ground. Here decomposers such as bacteria and fungi can digest the dead vegetation and pass this food on to small grubs, insects and eventually wood rats and owls. A simple food chain that exists in the redwood forest is

canadian redwood → wood rat → spotted owl

A **pyramid of energy** for this food chain is shown in Fig 3.1.

The savanna grasslands of Africa are filled with the noise of many animals. A food chain that exists here is

savanna grass → wildebeest → lion

The vast grasslands are grazed by wildebeest which can multiply to vast herds. Some wildebeest will be eaten by lions.

Exercise

> 1 Draw the pyramid of energy for the savanna food chain given above on a scale that compares with the redwood pyramid (Fig. 3.1).

Human population

The present-day bushmen of South Africa hunt their prey using bows, arrows and spears. Just as the eventual population of lions is limited by the availability of wildebeest food, so the population of bushmen is limited by how much hunted food is available to them. Both are top carnivores in their food chains and in South Africa there are just about as many bushmen as there are lions – they are both pretty rare.

Ancient humans were prolific hunters. Their population remained low because of their position as top carnivore in the food chain. Ten thousand years ago the world population of hunter–gatherer people was around five million – about half the present population of

London. At about this time the pattern of life for humans started to change. They discovered agriculture.

With agriculture, humans shifted their position in the food chain from top carnivore to herbivore and, just as the wildebeest herbivore population is very large on the plains of Africa, so the human population has increased explosively.

Deforestation and the loss of the land

The climate in the Sahara desert is hot and dry, but there used to be enough rain for crops, and animals were able to graze the scrubland. Since 1980 the rains have failed and by 1985 the countries surrounding the desert were in crisis with farm animals having died and food stocks exhausted. A century ago Ethiopia was half-covered with forest. The growing population felled the trees for firewood and extra farmland. This caused the two main parts of the disaster – a drop in the rainfall and soil erosion so that the Sahara desert expanded. Fig 3.2 shows the previous traditional farming system in Ethiopia and the situation today.

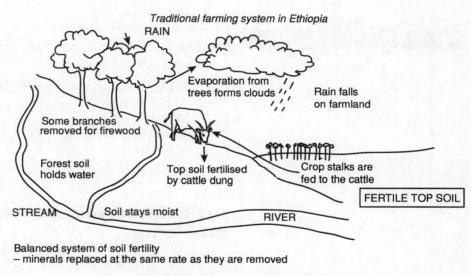

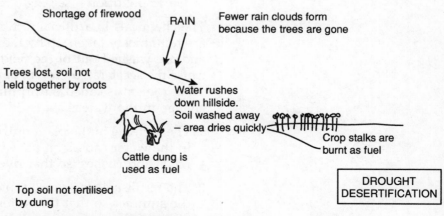

3.2   The traditional farming system in Ethiopia and (top) the situation today.

2 a In the traditional farming system in Ethiopia, how was the soil fertility maintained?

b Explain two ways in which the deforestation has caused the loss of the soil.

c Explain how the deforestation has contributed to the drop in rainfall in the area.

d What are the sources of fuel in Ethiopia now that there is no more firewood and what effect will the use of these fuels have on the fertility of the soil?

e Part of the water cycle is shown in Fig 3.2. Draw a diagram of the complete water cycle.

Poor countries such as Ethiopia cannot afford to buy artificial fertilisers to increase food production. However in richer parts of the world, especially since World War II, a 'green revolution' has occurred. New strains of cereals (needing larger quantities of fertiliser) which grow quickly and have high yields have been developed. Much of the artificial fertiliser put on to fields in this country is based on ammonium nitrate, NH_4NO_3. Plants absorb the nitrate into their roots. Small amounts of phosphate and potassium which are also needed by the plant, may also be added to the fertiliser mixture.

3 Use text books (in a library search) to answer the following questions.

a Why do all plants require nitrate?

b What symptoms in a crop plant may indicate a deficiency of nitrate in the soil?

c Why do plants, like all organisms, need phosphate (see Chapter 1)?

Artificial fertilisers – an environmental problem

Although we can see the benefits of fertilisers in increased crop yields, the use of artificial fertilisers on farm fields is an important aspect of water pollution – see Fig 3.3.

The chemicals in artificial fertilisers are soluble in water so if they are not quickly taken up by the crop plants, they dissolve in rainwater and are washed out of the fields into streams and rivers. Here, they cause the accelerated growth of pondweed and algae in the waterways. The problem is called **eutrophication** and it produces a step-by-step change in the waterway:

1 The increase in the plant nutrients (especially nitrate and phosphate) in the water causes an increase in the growth of plants, such as reeds and rushes, so that river channels can become choked.

2 The fertilisers also cause rapid growth of blue–green algae during the summer so that the water can become the colour of pea soup. Some of these organisms release poisons and make the water taste and smell foul.

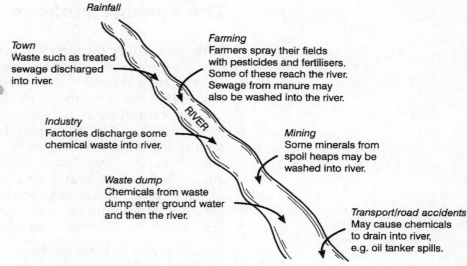

3.3   Different types of water pollution.

3 The dense colour of the water due to the growth of algae and pondweed mainly at the surface of the water shades large plants which grow on the bottom of the waterway. The consequent reduction of photosynthesis deep in the waterway means that less oxygen is available to dissolve in the water.

4 Most importantly, as the excess algae and other plants die, they are decomposed by aerobic bacteria which use up much of any oxygen left in the water.

5 Fish and invertebrates which are sensitive to low oxygen levels disappear while species which can tolerate low oxygen levels, like blood worms and *Tubifex* worms, become predominant.

Table 3.1 compares some features (in arbitrary units) of a normal lake and another lake suffering eutrophication.

Table 3.1

Feature	Season when measured	Normal lake	Eutrophic lake
Chlorophyll concentration	Summer	5–6	8–10
Chlorophyll concentration	Winter	2–3	3–4
Light penetration distance	Summer	4–6	1–1.5
Oxygen content	Summer	200–220	110–120

Exercise

4 a Explain the difference in chlorophyll concentration in the water of the normal lake from summer to winter.
    b Explain briefly how you could measure the amount of light penetration.
    c Explain how the figures for chlorophyll concentration and light penetration for the lakes in summer are related.
    d What chemical measurement could be used to indicate whether a lake was suffering from eutrophication?

The sewage problem

In the first half of the nineteenth century, untreated sewage ran through the gutters of London; rivers were stinking and water-borne diseases such as cholera and typhoid caused regular epidemics. By 1850 no fish could survive in the Thames and Parliament passed a bill to clean it up. A main drainage system was built for London which took the sewage further out into the Thames estuary. By the end of the century fish were returning to the river. However gradually, with further increased population and therefore more sewage, the river quality declined again and the river became virtually lifeless between 1950 and 1960. New sewage works were built in the Thames estuary during the 1960s and the water quality began to improve. In 1983 the first salmon for 150 years was taken from the Thames.

Sewage is rich in **biodegradable** organic material from human faeces. Decomposer organisms in the river consume this organic matter. They are aerobic microbes and use up large amounts of the oxygen dissolved in the water to create a **biological oxygen demand** (BOD). It is this depletion of oxygen that kills off the animal life, rather than any poisonous substances in the sewage. If the oxygen demand of the sewage is high enough, the water may lose all its oxygen. Aquatic life dies off and anaerobic decay predominates. The products of this decay include hydrogen sulphide which smells of bad eggs.

Sewage treatment

Most sewage works use a biological method of sewage treatment – most commonly a 'percolating filter bed' (Fig 3.4).

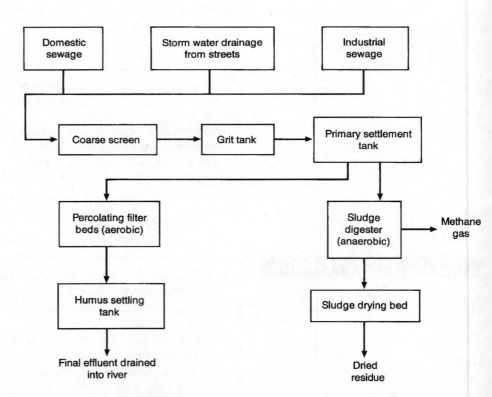

3.4   Layout of a sewage works.

The sewage is first allowed to settle in the primary settlement tanks. The sludge can be periodically drawn away for anaerobic treatment in the sludge digester.

The liquid from the primary settlement tank is sent to the percolating filters. These are large round tanks fed with liquid sewage from a rotating arm. The sewage percolates down through a bed of stones covered with a living film of aerobic microbes which consume the organic matter in the sewage. Nearly all the organic content of sewage is removed. The remainder is diluted and absorbed by naturally occurring microbes in a nearby river.

 **Exercise**

5 a Copy and complete the following table:

Part of sewage works	Description of the process that occurs
A Coarse screen	Removal of large objects that may be in sewage (e.g. disposable nappies)
B Primary settlement tank	
C Percolating filter beds	
D Humus tank	The liquid leaving the filter beds contains dead organisms that lived amongst the stones of the filter bed. This humus is allowed to settle to the bottom of the tank before the final liquid effluent is passed to a river.

b What is the purpose of all the stones in the deep percolating filter beds?

c Why are sewage works built on low lying land next to a river?

d The final effluent that is passed from the sewage works into the river is rich in nitrates and phosphates. What problem will these chemicals cause in the river?

e A stream receives some sewage from a farm. Why does the stream smell of rotten eggs during warm weather rather than cold?

The other method of biological treatment of sewage is the anaerobic digestion by microbes in septic tanks used in Britain for communities not connected to the mains sewage system. Rural farms in developing countries also put waste animal and vegetable matter into anaerobic digester tanks. When all this matter rots in the absence of air, a gas is given off. This **biogas** contains mainly methane which is a good fuel. The residual material from the bottom of the digester can be dried – it is rich in nitrates, phosphates and other minerals. Household rubbish tipped into landfill sites also produces biogas.

6 a Why is the gas called biogas?
  b What use can be made of the residue material from the bottom
     of the digester?
  c Why is biogas generation ideal for isolated farms and
     communities in the developing world?
  d What kinds of household rubbish would be best at producing
     biogas at landfill sites?

Biological indicator organisms of water quality

River authorities study the organisms present in a stream to reveal the
degree of pollution. Table 3.2 shows the indicator organisms used to
observe water quality.

Table 3.2

Pollution level	Biological indicator organisms
1 Clean water or very low pollution	Stonefly nymph, mayfly nymph
2 Slight pollution	Freshwater shrimp, caddis fly larvae
3 Moderate pollution	Blood worm, water louse
4 High pollution	Rat-tailed maggot, sludge worm
5 Heavy pollution	No animal life observed

Changes in the physical and chemical properties (especially the
amount of dissolved oxygen) in a river below the point of organic
discharge produce characteristic changes in the type and diversity of
aquatic organisms. Fig 3.5 displays some of the changes occurring
in the river downstream of the discharge point.

a What causes the drop in the amount of dissolved oxygen just after
   the point of effluent discharge?

   Answer: The aerobic microorganisms are using the organic matter as
   a food source, and then using much dissolved oxygen from
   the water to respire this organic 'food'.

b What is the likely cause of the rise in nitrate and phosphate
   concentration at the same point on the river?

   Answer: These minerals are released from the microbial breakdown
   of the organic matter.

c Which organisms from the graphs are the likely culprits of the
   changes noted in a and b?

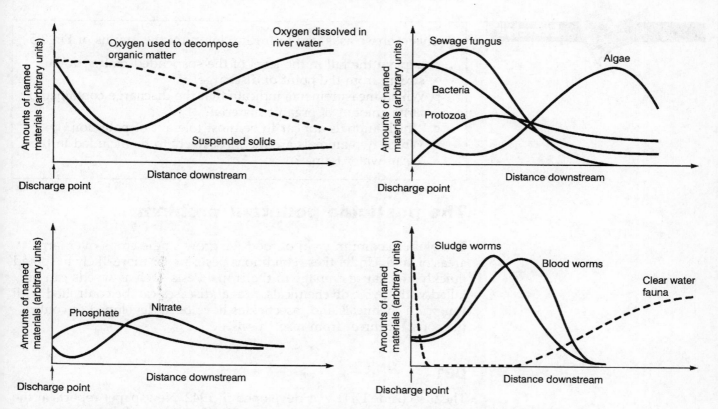

3.5   Some biological, physical and chemical changes occurring downstream of a sewage discharge point.

Answer: Sewage 'fungus', bacteria and protozoa.

d What is the relationship between the graphs for the changes in the algae population downstream and the gradual rise in oxygen concentration in the water?

Answer: As plants, algae carry out photosynthesis and produce oxygen. This means that the graph for the gradual rise in oxygen in the water will coincide with the graph for algae growth.

e The majority of the animals in the river are likely to become re-established after the algae and other plant populations have done so. Discuss the reasons for this.

Answer: Most of the animals (e.g. the clean water fauna) will need high concentrations of oxygen in the water and they will obtain their food down food chains. Both the oxygen and the food will be provided by the plant populations.

f Why would it be a good idea to install small weirs in a river downstream of any effluent discharge?

Answer: The turbulent water at a weir allows oxygen to dissolve from the air into the water. This raises the concentration of oxygen dissolved in the water that is so needed for the re-establishment of the animal life.

g Why would sewage discharged into a river in the summer months have a greater impact on the dissolved oxygen content of the water than the same discharge into the river in the winter?

Answer: The aerobic microorganisms are more active in warmer conditions and so reduce the oxygen levels faster during the summer months.

Exercise

7 This exercise asks you to further interpret the graphs of Fig 3.5.

 a Explain the fall in the level of the suspended matter down-stream from the point of discharge.

 b Which measurements indicate that the discharge contained a large amount of organic material?

 c Which animals appear to be most tolerant of pollution?

 d Which two animals are the most likely to be included in the 'clean water fauna'?

The pesticide pollutant problem

To obtain maximum yield of food we grow single crops over large areas of land. Under these conditions pests are far more likely to spread quickly and cause damage to the crops. Pests such as weeds can be killed with herbicide chemicals, fungal diseases can be controlled with fungicide chemicals and insecticides have been developed to control the constant threat from insect pests.

DDT

The insecticide DDT was developed in 1942. Newspaper reports at the time stated that this new chemical killed all insects but had no harmful effects on other animals. In 1952 it was reported that the use of DDT had nearly wiped out the mosquito that carried malaria and soon this disease would be eliminated. At the same time scientific journals were receiving accounts of the death of birds of prey such as eagles, ospreys and falcons in the USA and Europe.

DDT is not biodegradable. It does not break down easily and stays in the soil for many years. In animals' bodies it is not excreted, so that it accumulates in the tissues of any animal feeding on food contaminated with DDT. These organisms become the food for others which then accumulate even more. So the DDT becomes more concentrated as it is passed along the food chain as is shown in Fig 3.6.

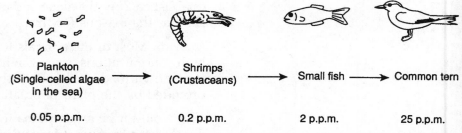

Plankton (Single-celled algae in the sea)	Shrimps (Crustaceans)	Small fish	Common tern
0.05 p.p.m.	0.2 p.p.m.	2 p.p.m.	25 p.p.m.

3.6   DDT accumulates along a food chain. (The numbers show the amount of DDT in parts per million in each kind of organism in the food chain.)

Exercise

8 During the 1960s the numbers of peregrine falcons in Britain declined. At this time cereal crops were often sprayed with insecticides such as DDT, Dieldrin and Aldrin. These substances can cause falcons to become infertile and also to lay thin-shelled eggs. Although the use of these insecticides is now restricted, large amounts of these chemicals have been washed into the oceans.

(The food web shown in Fig 2.1 may help you to answer the following questions.)

a Write out the food chain by which DDT sprayed onto a cereal crop could be passed to a peregrine falcon.

b Suggest how the peregrine falcon received large doses of DDT even though each seed on the cereal plant carried only a small amount of the insecticide.

c Suggest one way in which peregrine falcons could still be eating food containing DDT in the 1990s.

d Unlike DDT which is a very stable chemical, more recent insecticides are **biodegradable**. Explain the meaning of the term 'biodegradable' and describe the advantage for the consumer organisms in a food chain of using such insecticides on crops.

An alternative to using chemical pesticides is to use **biological control** which means using natural, biological methods to keep pest populations at a low level. The most usual method is to introduce a natural predator or parasite into the pest population. For example:

- A serious pest in greenhouses, the red spider mite, can be reduced considerably by the introduction of the predatory mite *Phytoseiulus persimmilis*.
- In butterfly farms the vegetation is prone to attack by insects called thrips. Insecticides would kill the butterflies as well as the thrips. A predatory mite is used to eat the thrips but not the butterflies.

One big advantage of biological control over the use of pesticides is that it is specific. The other advantage of biological control is that its effect can be long-lasting if a dynamic predator–prey relationship is set up.

Radiation

A modern food chain pollutant, which causes an ever-increasing problem to our society is the radiation from radioactive products and nuclear waste. Radiation damages the DNA genetic material of cells, causing genes to mutate and upsetting normal cell division so that it can induce cancers.

Radioactive waste is produced from industry, hospitals and nuclear reactors and it must be stored until the radioactivity has decayed. At present it is often buried deep in disused mines or dumped at sea.

In 1986 an explosion at the Chernobyl nuclear reactor sent a radioactive cloud into the atmosphere. Shortly afterwards radioactive rain fell on the mountains of Scandinavian countries and Britain. Many reindeer in Lappland had to be slaughtered because they had eaten radioactive vegetation and their meat and milk were no longer fit for human consumption. In Britain the food chain, grass → sheep → human was affected since radioactive fallout had rained on the hills of Wales and the Lake District. Radioactivity can last for many years and some sheep farmers could not sell their lambs for meat five years after the Chernobyl disaster.

Atmospheric pollution

In the 1950's many British cities suffered from smogs caused by the smoke from house and factory chimneys. Parliament passed the Clean Air Acts which prohibited the use of smoky fuels in built-up areas. Nowadays air pollution problems are more serious than smog because they affect the whole planet rather than just industrial areas. The major air pollution problems are acid rain, global warming from the greenhouse effect and depletion of the ozone layer.

Acid rain

Since 1950 sulphur dioxide from power stations and nitrogen oxide emissions from road traffic have doubled in Europe as a whole.

Three gases, normally present in the atmosphere in small amounts, dissolve in water vapour to give acidic solutions so that rain in the cleanest of atmospheres is slightly acidic with a pH of around 5.6. These gases are:

1 Carbon dioxide, from the respiration of all living organisms and forest fires, which dissolves in water to produce weak carbonic acid.

$$CO_2 + H_2O \rightarrow H_2CO_3$$

2 Nitric oxide produced by lightning storms dissolves to give nitric acid.

$$4\ NO + 3\ O_2 + 2\ H_2O \rightarrow 4\ HNO_3$$

3 Sulphur dioxide is produced by volcanoes and gives rain containing sulphuric acid.

$$2\ SO_2 + O_2 + 2\ H_2O \rightarrow 2\ H_2SO_4$$

During this century an increase in the sulphur dioxide concentration in the atmosphere has been caused by the burning of fossil fuels in power stations and industry. Also the high temperature of car engines causes nitrogen oxides to form from the nitrogen and oxygen in the air. As the amounts of these gases in the atmosphere have increased, then so has the problem of acid rain. Instead of unpolluted rain of pH 5.6, some areas can now experience very acidic rain of pH 2.4 – rain more acidic than vinegar.

Most of the pollutant gases released into the atmosphere (around two-thirds) do not dissolve in rainwater and are blown on the wind to return to earth as gases. This is called **dry deposition**. The remaining one-third dissolves in the water vapour of clouds and falls as acid rain or snow – **wet deposition**.

The effects of acid rain depend very much on the type of soil where the rain falls. If the soil is alkaline, as it is in limestone areas, then the acid rain will be neutralised.

Exercise

9 Limestone is calcium carbonate, $CaCO_3$. Write a word and balanced symbol equation for the neutralisation of nitric acid, HNO_3, by limestone, to form calcium nitrate $Ca(NO_3)_2$.

If the soil is thin and less alkaline, not only is the acid not neutralised but it runs off into the water courses. Such thin soils occur in the mountainous areas of Europe such as Wales, Scotland, Scandinavia and the Alps. The problem is made worse by the extensive planting of conifer forests in these areas. Recent research has shown that the conifer needles efficiently capture both the wet and dry deposition of acidic pollutants and concentrate the acid conditions in the forest soil.

Acid conditions in lakes and rivers reduce the reproduction rate of salmon and trout. The eggs fail to hatch because the enzyme produced by the larval fish to dissolve the egg membrane is inhibited by a reduction in pH.

Reduction in pH means increase in acidity

Acidity dissolves metal ions from the soil into the water courses. In alkaline soils these are likely to be the non-toxic metal ions of calcium or magnesium. But in other soils the aluminium ion may dissolve from the soil into the waterways and this is toxic to freshwater fish and invertebrates causing damage to their gills.

Thousands of lakes in Norway and Sweden have lost their fish to acid pollution. In Britain the problem seems to be less serious, but some rivers in mid Wales, the Lake District and Scotland have reported fish deaths, especially after the melting of overlying snow at the end of winter.

Exercise

10 The hatching of trout and salmon eggs coincides often with the period when snow is melting. Why will this have a deleterious effect on these fish populations?

The only real solution to acid rain is to reduce the emission of acid gases from the burning of fossil fuels. Sulphur dioxide scrubbers at power stations reduce emissions, but need large amounts of calcium carbonate (limestone) to operate. Similarly, to neutralise the acidity building up in a polluted lake, large amounts of limestone are again needed.

Exercise

11 Discuss what other environmental damage would be caused by the limestone treatment of acid pollution on this scale.

The greenhouse effect

The greenhouse effect and global warming seem to be the environmental issues of the 1990s. When the heat from the Sun reaches the atmosphere around the Earth, short-wave radiation passes through and warms the air and the Earth's surface, but most of the longer-wave radiation given off by this warming cannot pass back out through the atmosphere, so this heat is trapped. The atmosphere acts as a sort of 'one-way blanket', keeping the Earth much warmer than it would otherwise be. The same effect occurs in a parked car in summer or in a greenhouse. As a result of this 'greenhouse effect' the surface temperature of the Earth is kept at an average of 15 °C. Without the greenhouse effect of the atmosphere the Earth's surface temperature would be around −18 °C, too cold for life as we know it.

The greenhouse effect in itself is essential to life, however an increase in the atmosphere of so-called **greenhouse gases** may lead to intensification of this effect and **global warming**. Carbon dioxide contributes the largest amount to the greenhouse effect. Other gases like methane, ozone and chlorofluorocarbons (CFCs) contribute less because they are less abundant in the atmosphere, although they are, in fact, more efficient greenhouse gases.

There are two likely consequences of an increased greenhouse effect:

- Global warming might lead to a higher sea level because the water in the seas would expand and some of the polar icecaps would melt.
- Increased temperatures may also have serious effects on the Earth's climate, threatening farming. Some fertile areas may become deserts. Pests and diseases within an area may change dramatically. Crops grown may have to be altered to ones suitable for the new conditions.

The amount of carbon dioxide in the atmosphere has been rising since the industrial revolution as we have burnt more and more fossil fuels. Deforestation and the poisoning of phytoplankton in the sea by pollution makes the problem worse by reducing the amount of global photosynthesis. Fig 3.7 shows the present day estimates of the amounts of carbon moved around parts of the carbon cycle.

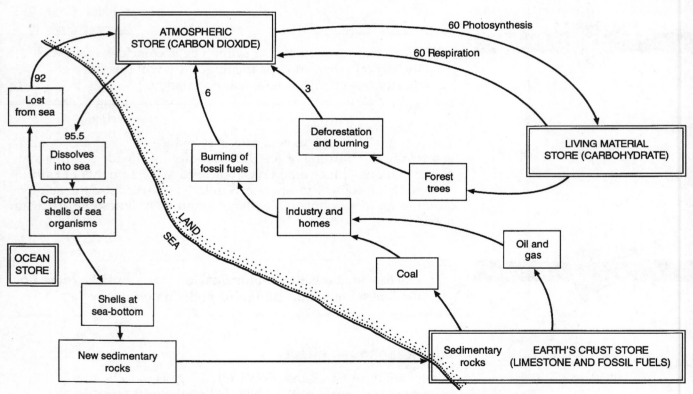

3.7 Amounts of carbon (in billions of tonnes per year) moved around parts of the carbon cycle.

Exercise

12 a In which of the four stores will the carbon remain the longest?
   b The Earth's crust store is mainly limestone. In what chemical form is the carbon in this store and how was the limestone formed originally?
   c Name the two stores that are losing carbon to the atmosphere due to human activity.

> d Over millions of years the carbon cycle has been in balance. Use the figures to show by how much, overall, the cycle is going out of balance.
>
> e Increased temperatures on land and in the sea, due to global warming, may increase the photosynthetic activity of plant material. What effect might this have on carbon dioxide levels in the atmosphere and global warming?

The ozone layer

Ozone, O_3, is found high up as a layer in the Earth's atmosphere. It is made by the action of ultra-violet (UV) radiation on oxygen, O_2, molecules. It acts as an atmospheric filter for the high amounts of ultra-violet radiation that reach the Earth from the Sun.

$$3\,O_2 \quad \rightarrow \quad 2\,O_3$$

$$\text{oxygen} \qquad \text{ozone}$$

Gases such as the nitrogen oxides and the chlorofluorocarbons (CFCs) act as catalysts to reverse this action, changing ozone back to oxygen.

Exercise

> 13 The CFCs are very unreactive. They act as **catalysts**. Why does this make the problem of getting rid of them very difficult?

In 1982 British Antarctic Survey scientists reported a hole in the ozone layer over the South Pole. Since then a further thinning of the layer has been observed over both poles. UV radiation can damage the DNA genetic material in cells and can therefore cause mutations and cancers. In humans it causes skin cancer and eye cataracts. It could harm plankton in the sea – the very base of the food chain. Research is being carried out on the effect of increased UV light on plants.

Ozone also causes difficulties closer to the Earth's surface, but here it is a problem of too much rather than too little. It is produced when gases in the exhaust fumes of cars react together in the presence of strong sunlight. Ozone can cause breathing difficulties for people suffering from asthma and it damages paint, plastics and dyes.

Exchange

A living cell has to exchange different materials with its environment (see Fig 1.5). The exchange of four materials is required just for respiration to occur. Food and oxygen must enter the cell whilst carbon dioxide and other waste products must leave. In a multicellular organism these materials have to be transported around the whole organism in order for each cell to exchange materials.

This chapter looks at the different mechanisms used for the exchange and transport of materials within the organism. The three main exchange processes used by cells are **diffusion, osmosis** and **active transport**.

Diffusion

Why does the smell of a cooked breakfast reach us in the bedroom? It is because gas molecules are in rapid random motion so they spread out from where they are concentrated to where they are less concentrated. This diffusion of molecules occurs whenever a concentration gradient exists and continues until eventually the molecules are evenly distributed throughout and an equilibrium is reached.

Diffusion is important for the movement of molecules in and out of cells. Oxygen inside a cell is continually being used for respiration so that there will be less oxygen inside the cell than in the surrounding tissues. This concentration gradient results in oxygen molecules continually diffusing into the cell from outside. The reverse is true for carbon dioxide. This is produced in the cell by respiration so that there will always be more of it in the cell than outside. Its concentration gradient, therefore, ensures that it will continually diffuse out of the cell.

In multicellular organisms, diffusion also occurs in organs as well as in cells. Specialised organs exist in order to exchange materials between the outside environment and the body of the organism. These materials are then transported around the body to all the cells. In most animals it is the blood system that carries out this transport.

Some organs specialised for exchange of materials are the lungs or gills, gut and kidney in animals and the leaf and root in plants.

Fig 4.1 shows diagrams of a leaf section, an alveolus from the lungs and a villus from the small intestine of a mammal. Labelled arrows on the leaf section show the overall diffusion movements down the concentration gradients for different substances.

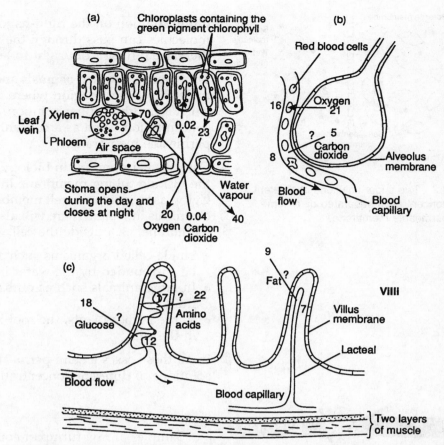

4.1 Sections of (a) a leaf, (b) an alveolus and (c) villi. (The numbers show the relative concentrations of the named substances.)

Exercise

1 Copy Fig 4.1 and complete the lines with a question mark to show the direction of diffusion movements for the alveolus and the villi.

2 Apart from the necessary concentration gradient, give two other adaptations for exchange that are present in both the alveolus and villus.

For diffusion to occur the membranes of cells or tissues must be fully permeable to the molecules in question. The cell membrane, however, is not permeable to all substances. The membrane has pores and only molecules small enough to pass through them will diffuse through. The cell membrane then, is **semi-permeable**, allowing the diffusion of small molecules but not larger ones.

Another term for semi-permeable is **selectively permeable**, or **differentially permeable**.

Osmosis

Fig 4.2 shows a concentrated sugar solution separated from a dilute sugar solution by visking tubing – a semi-permeable membrane. There is a higher concentration of sugar molecules on the left-hand side of the membrane so that sugar molecules would normally diffuse from the left to the right. This cannot happen, however, because the pores in the membrane are too small for the large sugar molecules to pass through. A concentration gradient also exists for the smaller water molecules. There is a higher concentration of these in the more

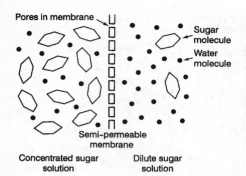

4.2   Two sugar solutions of different concentration separated by a semi-permeable membrane.

dilute solution on the right-hand side of the membrane. The water molecules can pass through the pores in the membrane and will therefore diffuse from right to left until an equilibrium is reached.

This process is called **osmosis** and it is defined as the diffusion of water molecules from a region where they are in a higher concentration (such as a dilute sugar solution) to a region where they are in a lower concentration (such as a concentrated sugar solution) through a semi-permeable membrane.

Osmosis is important in biology because cell membranes behave like the visking tubing membrane in Fig 4.2 – they are semi-permeable. Cytoplasm, inside the cell membrane, is a complex solution of different chemicals in water. There will also be a solution on the other side of the membrane, i.e. outside the cell.

- Single-celled organisms such as *Amoeba* and many bacteria will be surrounded by the water in which they live.
- In large animals such as ourselves, cells are surrounded by tissue fluid.
- In the higher plants, the root hair cells grow into the film of water in the soil.

In all these cases a semi-permeable cell membrane separates two solutions of different concentration so osmosis will occur.

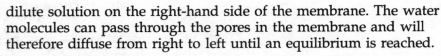

3 Samples of raw turnip of equal mass are weighed, and soaked in a range of different sugar solutions and also in distilled water. After an hour, the turnip pieces are blotted dry, and reweighed. The table opposite shows the results.

   a Work out the loss or gain in mass of the turnip sample in each liquid.
   b Plot your results on a graph of loss or gain (vertically) against concentration of sugar (horizontally). Gains should go *above* the horizontal axis and losses *below.*
   c At what concentration of sugar does no change in mass take place?
   d What concentration of sugar is equal to the internal concentration of the turnip cells? How did you get your answer?
   e What loss or gain in mass would occur if you soaked some turnip in
      i   0.32 mol/l sugar solution
      ii 0.85 mol/l sugar solution?
      How did you get your answers?

Concentration of sugar (mol/l)	Original mass of turnip sample (g)	Final mass of turnip sample (g)
distilled water	8.5	13.3
0.1	8.5	13.0
0.2	8.5	11.6
0.3	8.5	9.5
0.4	8.5	9.1
0.5	8.5	8.6
0.6	8.5	8.0
0.7	8.5	7.7
0.8	8.5	6.3
0.9	8.5	5.2
1.0	8.5	5.1

Water potential

In order to show how much osmosis will occur in a cell the term **water potential** (Ψ) is used. This is a measure of the ability of a cell or a solution to *lose* water through osmosis. The numbers used for water potentials may seem a little strange. Pure water has a water potential of zero. This is the highest number because pure water has the greatest ability to lose water by osmosis. More concentrated solutions will have less ability to lose water and therefore have lower water potentials – negative number water potentials. All Ψ values are measured in pressure units of megapascals (MPa) or kilopascals (kPa). Look at the list below to see how it works.

Ψ is a Greek letter psi, pronounced 'sigh'.

0.0 MPa   Water potential of pure water (highest Ψ possible)
− 0.2 MPa
− 0.4 MPa   Water potential of a dilute solution (low Ψ)
− 0.6 MPa
− 0.8 MPa   Water potential of a concentrated solution (lower Ψ)
− 1.0 MPa
− 1.2 MPa   Water potential of a still more concentrated solution

Follow the worked example to check you understand this.

Example

During osmosis, which way will water move for the following solutions, A and B, separated by a semi-permeable membrane?

$$\begin{array}{c|c} A & B \\ \Psi = -0.8 \text{ MPa} & \Psi = -0.4 \text{ MPa} \end{array}$$

Semi-permeable membrane

Answer: Water will move from the solution with the higher water potential to the solution with the lower water potential until there is a balance (or equilibrium). − 0.4 is a higher number than − 0.8, so water will move from B to A until equilibrium is reached when each solution has a water potential of − 0.6 MPa.

Exercise

4 Fig 4.3 shows a root hair cell penetrating the water film around soil particles. The water potentials of the soil water and the root hair cell are indicated.

a Give a reason for the soil water having a water potential of − 0.1 MPa rather than 0.0 MPa for pure water.
b Use the water potentials to show in which direction water will move.

5 Fig 4.4 shows two adjacent cells (A and B) with their corresponding water potentials.

a Which way will water flow?
b Give the value of the water potential of the cells when equilibrium is reached.

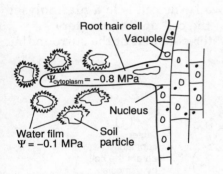

Root hair cell
Vacuole
$\Psi_{cytoplasm} = -0.8$ MPa
Nucleus
Water film
Ψ = −0.1 MPa
Soil particle

4.3 A root hair cell penetrating the surrounding soil water.

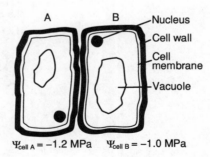

A   B — Nucleus
— Cell wall
— Cell membrane
— Vacuole

$\Psi_{cell\,A} = -1.2$ MPa $\Psi_{cell\,B} = -1.0$ MPa

4.4 Two adjacent plant cells.

Osmosis and plant cells

Unlike animal cells, plant cells are surrounded by a cell wall. This is fully permeable to most molecules. It is also very strong and will only let the cell expand a little (think of a bicycle tyre wall). Just inside the cell wall is the semi-permeable cell membrane (the inner tube of the tyre).

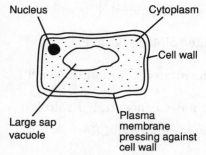

4.5   Plant cell in pure water.

Fig 4.5 shows a plant cell in pure water. The cell will take in water by osmosis because it will have a lower water potential than the pure water outside. This causes the cytoplasm and the vacuole to swell. The cytoplasm presses out against the wall but this soon resists and presses back on the cell contents. Water will continue to enter the cell by osmosis until the pressure from the cell wall is equal to the pressure of water entering the cell.

In this state the plant cell is like a fully inflated tyre – firm and slightly swollen. It is said to be **turgid** and the rigidity of the cell provides support to plant tissues.

Fig 4.6 shows a plant cell bathed in a concentrated solution with a lower water potential than the cell. It will lose water by osmosis and the cytoplasm and the vacuole will shrink. The cytoplasm will stop pushing outwards on the cell wall, and, like a tyre when some air is let out, the cell will become floppy or **flaccid**. Such a cell offers no support to the surrounding tissues and wilting results.

6 In terms of osmosis, describe the consequences of the flooding of garden plants with sea water.

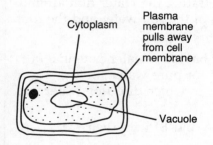

4.6   Plant cell in a concentrated solution.

In a more concentrated solution, more water will diffuse out of the cell and the cell contents will shrink to the extent that the cell membrane pulls away from the cell wall. The cell is in a **plasmolysed** state. Water continues to move out of the cell until the water potential of the cell is equal to the water potential of the surrounding solution.

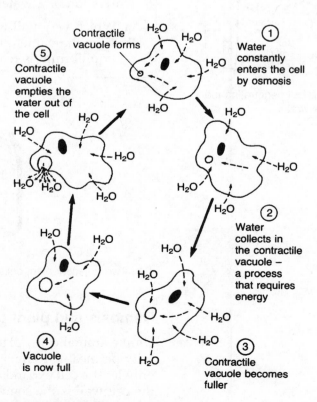

4.7   Osmoregulation in *Amoeba*.

Osmosis and animal cells

Animal cells do not have cell walls. In pure water an animal cell would take in water by osmosis and begin to swell. The delicate cell membrane would stretch only a little before the cell bursts. Animal cells, then, are a lot more fragile than plant cells, so water and solutes need to be regulated carefully. This is called **osmoregulation** and one way of achieving it is through **active transport** of materials.

Single-celled freshwater organisms such as *Amoeba* and *Paramecium* have lower water potentials (due to the chemical solution in their cells) than their surrounding watery medium. Being single-celled this poses a survival problem: water flows into their cells by osmosis and if this went unregulated the cell would burst. However, these organisms possess a mechanism for surviving in these conditions: they have one or more osmoregulatory contractile vacuoles. These act like bilge pumps on a boat and rid the cell of excess water. Follow through Fig 4.7 showing how the contractile vacuole of *Amoeba* works. Note that the excess water entering the cell is actively transported into the vacuole before it is expelled from the cell.

Active transport

Diffusion is a purely physical process in which molecules move from a region of higher concentration to a region of lower concentration. There are, however, some biological situations where substances are seen to move from a region of low concentration to one of higher concentration. The contractile vacuole of *Amoeba*, for example, contains almost pure water and yet more water enters it from the cytoplasm of the cell which is at a lower water potential. By osmosis, water would move in the opposite direction. This movement of a molecule or ion against the concentration gradient requires energy from the cell and is called **active transport**.

Cells that carry out active transport on a large scale have an unusually large number of energy-supplying mitochondria. It is thought that a carrier substance in the cell membrane attaches itself to a molecule on the outside of the membrane and carries it to the inner surface where it is unloaded into the cytoplasm. It then returns to the outer surface to repeat the process. Evidence points to the need for energy for this return journey.

Active transport occurs throughout living systems. Some other examples are listed below.

- Active transport occurs in the phloem of the vascular bundles of plants where some materials move up in the phloem and others move in the opposite direction.
- Nerve axons use active transport to pump out sodium ions and so maintain an electric potential difference across the membrane of the neurone – the basis for the operation of the nerve impulse.
- The pacemaker of the heart also uses active transport to pump ions across cell membranes and so maintain a 'clock' mechanism to regulate the heart contractions.
- Plant roots absorb some minerals against the concentration gradient by active transport.
- Active transport is involved in kidney function and helps achieve osmoregulation of the blood, i.e. to keep the blood at a constant concentration of water and solutes.

4.8 Water content in a typical human.

A further factor in osmoregulation is the **hydrostatic pressure** caused by the flow of blood through the various blood vessels.

Hydrostatic pressure

Hydrostatic pressure in the blood

Blood circulates in our bodies under varying pressure from the pumping action of the heart. The blood travels in arteries, veins and capillaries that change in diameter and lumen (Fig 4.9). The elasticity of their walls causes more changes in pressure.

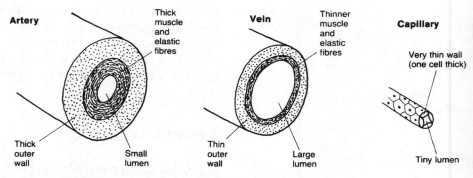

4.9 Cross sections through an artery, a vein and a capillary.

7 Do you understand the connection between **blood**, **plasma** and **tissue fluid**? Write a brief description of what you think the connection is, and then check the glossary at the back of the book.

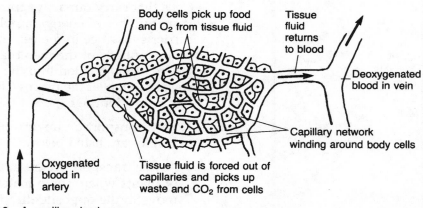

4.10 A capillary bed.

Although tissue fluid is mostly water, it does not leave the blood by osmosis. It is forced out under a pressure called a hydrostatic pressure, which is the pressure that develops when blood enters a capillary bed as in Fig 4.10.

8 Which part of the capillary bed has the greatest hydrostatic pressure?

As the tissue fluid bathes the body tissues, osmosis will occur to maintain the correct balance of water in the cells. (Nutrients and wastes will also diffuse in and out of the active cells.)

Remember that blood contains both water and dissolved solutes. This means there will be an overall water and osmotic potential in the blood opposing the hydrostatic pressure, and so drawing water back into the capillaries.

> 9 Where in a capillary bed will most fluid be returned? Briefly explain your answer.

Hydrostatic pressure in plants

Hydrostatic pressure is also involved in the fast movements of the leaves of the Venus fly trap and certain Mimosa plants. Hinge cells in these structures are under pressure since they are filled with water. Release of some of the water from the cells triggers a sudden drop in pressure which causes the fast movement.

A hydrostatic pressure is involved in the transpiration movement of water from the roots to the leaves of plants. Water enters the root hair cells by osmosis. It eventually enters the xylem vessels and is drawn up the tubes of this tissue to the leaves where it is required for photosynthesis. Weak attractive forces link water molecules together so that when water evaporates from the stomata of the leaves, more water molecules are drawn into their place. This is rather like railway trucks linked together; movement of the first truck will pull the whole train of trucks along.

The force of pulling water up the trunk of a tree is sufficient to cause the trunk circumference to change as shown in the graph of Fig 4.11.

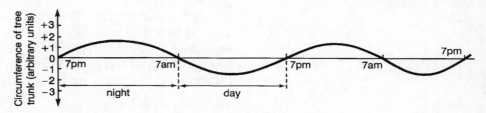

4.11   Changes in the circumference of a tree trunk over a 48-hour period.

During the day there is more evaporation of water from the leaves than at night because the stomata are more fully open and the daytime temperature is higher. This greater transpiration pull during the daylight hours causes the trunk circumference to be smaller during the day than at night.

> 10 The graph shown in Fig 4.11 represents a summertime period of moderate day and night temperatures. Sketch the graph during a period of similar night-time temperatures but much warmer days, and explain why its shape has changed.

Hydrostatic pressure in the kidneys

There is a particularly high hydrostatic pressure in blood entering the capillaries of the nephron, the working unit of the kidney. Fig 4.12 shows a diagram of a nephron.

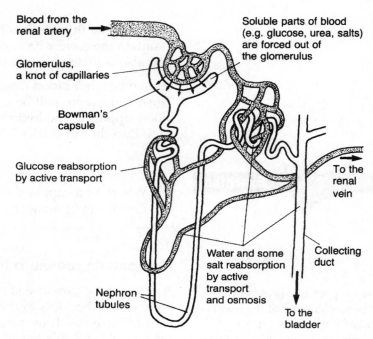

4.12   A nephron.

Exercise

11 What do you notice about the blood entering the glomerulus
    and the blood leaving it that contributes to this high pressure?

Because of the high hydrostatic pressure of the blood in the
glomerulus, small molecules such as water, urea, salt and glucose
are forced out into the Bowman's capsule of the nephron. Molecules
such as protein antibodies and the cells of the blood are too large to be
squeezed out through the small pores in the membrane between the
glomerulus and the capsule and so stay in the blood. This pressurised
filtration of the blood by the kidneys is called **ultrafiltration**.

The body needs to get rid of the waste product urea and excess water
and salt. These materials pass on through the nephron and will
eventually make up the main components of urine. Glucose is vital to
the body as the fuel for respiration, but is squeezed out of the
glomerulus blood simply because it is a relatively small molecule.
Glucose must be reabsorbed from the nephron filtrate back into the
blood or it will be lost in the urine. This reabsorption occurs in the first
twists of the nephron after the Bowman's capsule. Reabsorption occurs
against the concentration gradient and is achieved by active transport.
Before leaving the kidney the water and salt content of
the blood are adjusted by some reabsorption of them back into the
blood using both active transport and osmosis.

Dialysis and the artificial kidney

If the kidneys fail to work properly the blood will retain urea, water
and salt which are normally excreted in urine.

Urea is made in the liver from excess protein in our diet. We digest
protein in our gut to its constituent amino acids which then diffuse
into the bloodstream. Amino acids will be used for the immediate
growth and repair of the body by being rebuilt into the proteins that
help make the structure of the body. Excess amino acids are not stored,

R
|
NH₂ — C — C ═ O
| \
H   OH

Amino group → (points to NH₂)

Carboxylic acid group ← (points to C═O / OH)

Deamination removes
the amino group

Different 'R' for different amino acids
R = H for glycine
R = CH₃ for alanine

This residue contains carbon,
hydrogen and oxygen and can be respired.

NH_3 = Ammonia – waste product of freshwater fish
NH_2CONH_2 = Urea — waste product of mammals
Uric acid — waste product of birds and reptiles

4.13   Deamination of an amino acid in the liver.

they are **deaminated** in the liver (Fig 4.13). The nitrogen-containing amino group is stripped from the amino acid leaving a residue molecule which can be respired. The liver converts the toxic amino residues into less-toxic urea which is released into the blood and later removed by the kidneys. Urea is a small molecule, formula NH_2CONH_2. If it were to remain in the blood it would exert its toxic qualities and the water potential of the blood would be reduced.

Exercise

12 What would be the effect of blood with a reduced water potential on the surrounding tissue cells?

Water forms a large part of the diet and is also made in the body as a product of respiration. All cells need an adequate supply of water, but excess is excreted from the body. If this excretion did not occur, the blood volume and blood pressure would increase and the water potential of the blood would be disturbed.

Exercise

13 Would this retention of water cause the blood water potential to increase or decrease? Give a reason for your answer.

Salt is another component of the diet. It is required for healthy nerve and muscle function. Excess is excreted daily and, if this does not occur, nerve and muscle function and the water potential of the blood are disturbed.

Exercise

14 Would this retention of salt cause the blood water potential to increase or decrease? Give a reason for your answer.

Dialysis

If a person's kidneys stop working, illness and death follow soon after. The alternative to a kidney transplant is dialysis treatment using an artificial kidney machine. The machine has to be used three times a week to keep a patient alive.

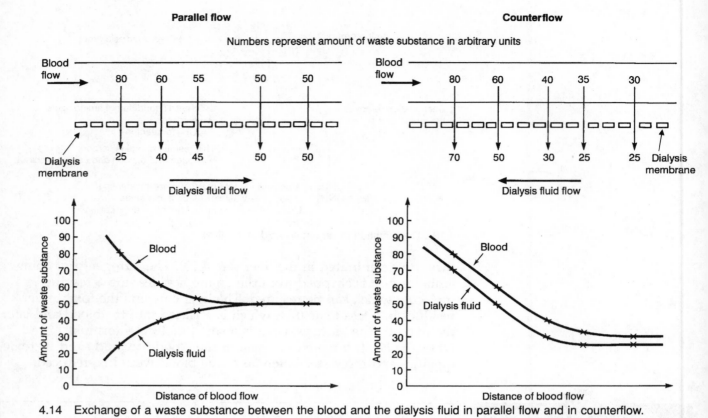

4.14   Exchange of a waste substance between the blood and the dialysis fluid in parallel flow and in counterflow.

Dialysis is similar to osmosis in that it involves the diffusion of small molecules through a semi-permeable membrane. While osmosis involves the movement of water molecules only, dialysis is designed to allow the diffusion of the small waste molecules (described above) out of the blood.

In the artificial kidney the patient's blood is separated by a semi-permeable membrane (similar to visking tubing) from the dialysis fluid which collects the waste molecules. Fig 4.14 shows the exchange of a substance that occurs when the blood and dialysis fluid are in either **parallel flow** or **counterflow** to each other.

15   Use the graphs to explain which type of flow gives the most efficient exchange of the substance from the blood into the dialysis fluid.

Counterflow mechanisms are also found in the living kidney for the re-absorption of some water from the urine back into the blood and in the gills of fish where oxygen and carbon dioxide are exchanged between the fish's blood and the surrounding water.

Although the membrane of the artificial kidney has tiny pores, it has proved impossible to achieve a pore size through which only the waste products escape. So glucose molecules are likely to move through the membrane and be lost from the blood.

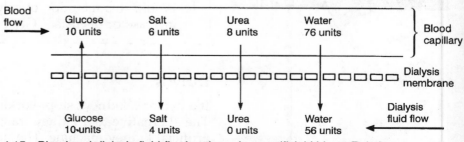

4.15   Blood and dialysis fluid flowing through an artificial kidney. Relative concentrations are given for each substance in the patient's blood and the dialysis fluid

Control over which substances are lost from the blood during dialysis is achieved by regulating the recipe of substances put into the dialysis fluid. Fig 4.15 shows the blood and the dialysis fluid. The relative concentrations of four substances in both fluids are displayed.

Use the information in Fig 4.15 to work out the diffusion movements of   a glucose   b salt.

a **Answer:** The glucose concentrations are the same in the patient's blood and the dialysis fluid. There is no concentration gradient and, although the glucose molecule can pass through the semi-permeable membrane, no overall movement of glucose from the blood will occur.

b **Answer:** The concentration of salt is slightly higher in the blood than in the dialysis fluid so that, overall, some salt will move down the concentration gradient from the blood to the dialysis fluid. An equilibrium will be reached when the blood ends up with the same salt concentration as the dialysis fluid.

16 Use the information in Fig 4.15 to work out the diffusion movements of   a urea   b water.

Gas pressures during breathing

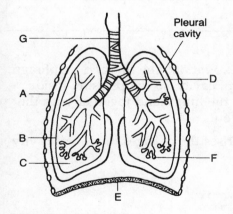

4.16   The chest cavity.

Our cells constantly exchange oxygen for carbon dioxide during respiration. As fast as the oxygen is used, the supply must be renewed and the waste gas removed. For small animals the process of diffusion is sufficient for this exchange to occur. Earthworms, for example, have thin, moist skins through which the gases can diffuse. Insects have air-filled tubes (tracheoles) throughout their bodies which allow the gases to diffuse from the air to deep into their tissues. For larger insects, such as a locust, this diffusion is not quite sufficient and the insect has to pump its body in order to increase the rate of gas exchange through the tracheoles.

For even larger animals, so much gas exchange is required that specialised organs (gills and lungs) have evolved that allow increased amounts of gas diffusion through their very large surface areas. The breathing movements of the **respiratory cycle** in animals with lungs are caused by differences in gas pressures.

17 a You should be able to recall the details of the structure of the lungs and the route air takes to get inside them. Place the following stages of that route in the correct order, starting with air inside.

air inside; bronchiole; alveolus; trachea; bronchus

b Copy and complete the labelling of Fig 4.16.

Pressures during the respiratory cycle

The important pressures inside the thorax during the respiratory cycle are those in the lungs and in the pleural cavity. Follow through the sequence of the respiratory cycle using Fig 4.17.

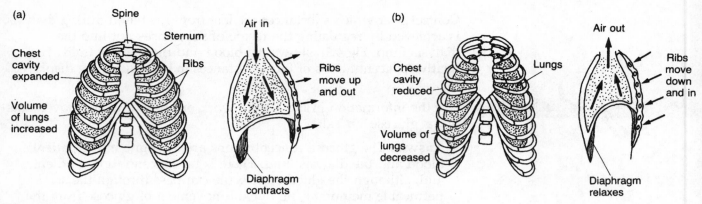

4.17   (a) Thorax during inspiration. (b) Thorax during expiration.

Immediately after breathing out (expiration), the thorax is at rest. Air pressure in the lungs is equal to atmospheric pressure – about 100 kPa – although the pleural cavity pressure is slightly less. This is because the elastic, stretchy lung tissue pulls away a little from the walls of the thorax.

Fig 4.17 (a) shows the thorax during inspiration (breathing in). The chest cavity has clearly expanded and this will lower the pleural cavity pressure further. Lung pressure drops to below atmospheric and air can rush in.

18 What effect will this have on the volume and pressure of the lungs?

Fig 4.17 (b) shows the thorax after expiration. The ribs and diaphragm move down and press on the pleural cavity, so the pressure inside rises. This increases the pressure in the lungs, and together with the elastic recoil of the lungs forces air out.

The graph below (Fig 4.18) summarises the changes in pleural cavity and lung pressures in an adult.

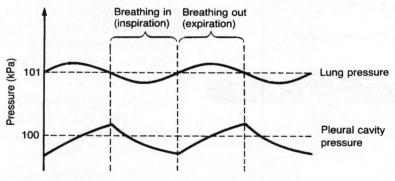

4.18   Graph showing changes in pleural cavity and lung pressures during the respiratory cycle.

Chapter 5

Control and coordination

How do animals respond to change?

STARTING POINTS

- Do you understand the following terms: **receptor, effector, nervous system, endocrine system**? Write a sentence or two (no more) to explain what each term means, then check your answer against the glossary at the back of the book.

All living organisms have the ability to respond to stimuli. A stimulus can come from inside the organism (internal stimulus), or from the environment outside the organism (external stimulus). The ability to respond to these stimuli is vital to the survival of every living organism.

Communication between the receptors, which detect stimuli, and the effectors, which respond to the stimuli, is brought about in two ways, by the **nervous system** and by the **endocrine system**.

The nervous system

Fig 5.1 is a diagram of the human nervous system. It consists of two main parts:

1 The central nervous system (CNS), which is made up of the brain and spinal cord.
2 The peripheral nervous system (PNS), which is made up of all the nerves which connect the body to the central nervous system.

Information from the environment is picked up by our receptors. Table 5.1 lists some of our receptors and the stimuli they detect.

Table 5.1 Receptors and stimuli

Receptor	**External stimulus detected**
Eye	Light
Ear (cochlea)	Sound
Ear (semi-circular canals)	Balance
Tongue and nose	Taste and smell
Skin	
Touch receptors (Meissner's Corpuscles)	Touch
Pressure receptors (Pacinian Corpuscles)	Pressure
Pain receptors (all over the body)	Pain
Temperature receptors	Heat and cold
Receptor	**Internal stimulus detected**
Pain receptors	Pain
Chemoreceptors	CO_2 levels in the blood
Baroreceptors	Blood pressure

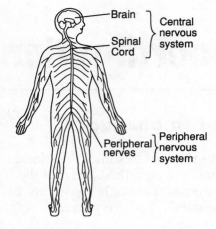

5.1   The nervous system

Nerve cells (neurones)

Linking the receptors, the CNS and the effectors, are nerve cells called **neurones**. There are three types of neurone – sensory neurones, relay neurones and motor neurones. Information from the receptors is relayed by sensory neurones to the spinal cord. From here, the information is passed on to the brain by relay neurones. When the brain has processed the information, other relay neurones carry the information back through the spinal cord to the appropriate place and pass the information on to motor neurones, which carry the information to the effector (an organ or muscle). The effector then carries out the response. Information is carried along the neurones as tiny electrical impulses which travel along the neurone at a speed of approximately 120 m/s.

Fig 5.2 summarises the path of the nerve impulse from receptor to effector. Because the brain was involved in deciding to move the arm, i.e. the response involved conscious thought, this response is said to be **voluntary**.

---

1 Look at Fig 5.1. If the spinal cord were damaged at the neck, what effects do you think this would have on the functions of the body?

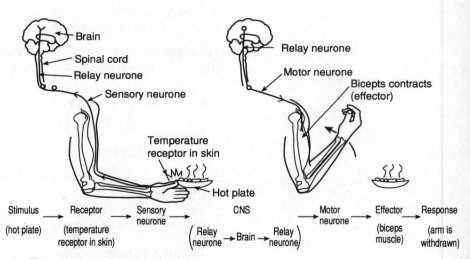

5.2   The pathway of a nerve impulse.

The three types of neurone differ slightly in appearance, but they all share the following common features:

- a cell body, which consists of a nucleus, cytoplasm and a cell membrane.
- an axon, a long thin extension of the cell body, filled with axoplasm. It can carry an impulse over long or short distances depending on its length. The axon is surrounded by a layer of fatty material called the myelin sheath, which is produced by special cells called Schwann cells. The sheath is a type of electrical insulator and helps to speed up the transmission of an impulse.
- dendrites, which are branching fibres that pass on an impulse from a sensory neurone to the relay neurones and pick up information from the relay neurones for the motor neurone.

Fig 5.3 shows the structure of (a) a sensory neurone, (b) a motor neurone.

Exercise

2 Look at Fig 5.3 (a) and (b). List the similarities and differences between a sensory neurone and a motor neurone in terms of structure and the direction of the impulse.

Reflex action

If a stimulus is particularly unpleasant, we have the ability to respond very quickly without conscious thought. This type of response is said to be **involuntary**, or reflex action. For example, if the plate in Fig 5.2 had been unbearably hot, the person would have snatched away the hand rather than thought about putting it down. Fig 5.4 shows the path taken by a nerve impulse in this case. When the impulse reaches the spinal cord via the sensory neurone, rather than going on to the brain, it travels through a relay neurone in the spinal cord straight to a motor neurone which carries the impulse directly to the biceps in the arm. This results in the muscle contracting, pulling the arm away. The reflex is a way of responding very quickly, in this case to a potentially harmful stimulus, without taking the time to 'consult' the brain. This reflex is located in the region of the spine at the shoulder and is therefore called a spinal reflex.

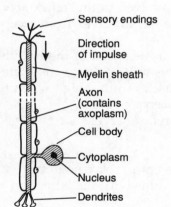

5.3   (a) A sensory neurone.

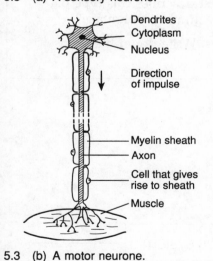

5.3   (b) A motor neurone.

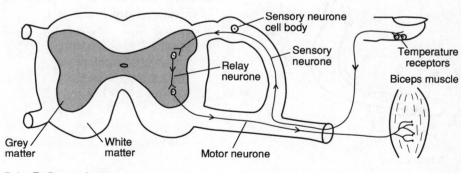

5.4   Reflex action.

The diameter of the pupil in the eye is constantly being adjusted, according to how much light is entering the eye. This is another example of a reflex. If the light is too bright, the pupil becomes smaller (constricts). If the light is too dim, the pupil becomes larger (dilates). Because light intensities are constantly changing, it would be wasteful if the brain had to be consulted every time. Because the neurones of the eye go directly to the brain, this is not a spinal reflex.

Exercise

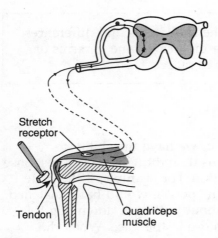

Stretch receptor

Tendon   Quadriceps muscle

5.5   The pathway involved in the knee jerk reflex

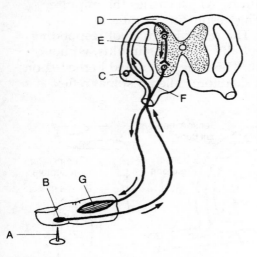

5.6   A reflex arc.

3 Fig 5.5 shows the reflex involved in the knee jerk reaction. When the tendon below the kneecap is tapped, an involuntary reaction follows, where the leg is quickly straightened. The knee jerk reaction prevents damage to the tendon when such pressure could over-stretch it. Study the diagram carefully and answer the following questions.

   a What is the structure that is stimulated by tapping with the hammer?
   b Outline the path taken by the nerve impulse from receptor to effector.
   c Which part of the spinal cord is involved in this reflex action?
   d Estimate the distance this impulse has to travel, from receptor to effector.
   e Based on your answer to part d, if the speed of the impulse was 110 m/s, how long would this reflex take to complete?
   f Describe what happens to the muscles in the thigh when the nerve impulse reaches the quadriceps.
   g What purpose do you think this reflex serves?

4 Fig 5.6 shows another reflex action.

   a Match effector; motor neurone; receptor; relay neurone; sensory neurone; spinal cord; stimulus with the letters on Fig. 5.6.
   b What would be the response in this reflex action?
   c Which of the following terms best describes this type of response: conscious, learned, involuntary, voluntary?

5 Table 5.2 summarises some reflexes. Write down the terms missing from boxes a–i.

What the reflex action tells us about the nervous system

Doctors often test the knee jerk reaction when they are trying to diagnose whether a patient has nerve damage caused by a bulging disc in the spine. Because this is a reflex, the brain is not involved and so the patient cannot fake the result of the test. If the response is missing, there is almost certain to be some damage to the spinal cord. Similarly, if somebody was involved in an accident and was unconscious, the doctor would open the patient's eye and shine a bright light into the pupil. Because the pupil reflex is not a spinal one, if it was absent, the doctor would immediately know if there was damage to the neurones in the brain.

Exercise

Table 5.2   Reflexes in muscle

Reflex	Stimulus	Receptor	Effector	Response
Secretion	Smell, sight and/or thought of food	Receptor cells in nose and eyes	Salivary glands	Saliva is produced
Blinking	Bright light	Retina or touch	Eyelid muscle	a
Knee jerk	b	Stretch receptors in muscle	c	Leg straightens
Movement away from pain	d	Pain receptors in skin	e	Foot lifted off sharp object
Movement away from heat	Picking up a hot test tube	f	g	Test tube is dropped
Coughing	Obstruction in the trachea	Receptors in the trachea	h	i

Look carefully at the diagrams of the paths followed by nerve impulses. There is a slight gap between the end of one neurone and the beginning of the next. This gap is called the **synapse**. We know that the nerve impulse is electrical and since electricity cannot jump across such a gap, how does it manage to get to the next neurone? The answer lies in the ability of the end of every neurone to secrete a chemical which diffuses across the synapse and stimulates the next neurone to set up another electrical impulse. The chemical is called acetylcholine (ACH). The arrival of a nerve impulse at the end of a neurone stimulates its release. It quickly diffuses across the synapse, binds with special receptors on the membrane of the next neurone and is quickly broken down. Its destruction is brought about by an enzyme called acetylcholinesterase, which is produced by the neurone it stimulates. Fig 5.7 summarises the events that take place at the synapse when a nerve impulse arrives.

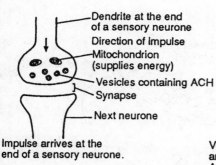

Impulse arrives at the end of a sensory neurone.

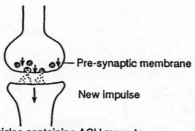

Vesicles containing ACH move to and fuse with the membrane releasing ACH into the synapse. ACH diffuses across the synapse and activates a new impulse in the next neurone (post-synaptic membrane).

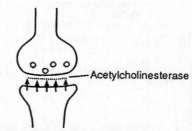

Post-synaptic membrane secretes the enzyme acetylcholinesterase, which destroys acetylcholine (ACH).

5.7   Events at the synapse.

Exercise

6　Can you think of a reason why it would not be desirable for acetylcholine to remain present in the synapse once it has stimulated the next neurone?

7　Fig 5.7 shows that mitochondria are found in abundance close to the vesicles that contain acetylcholine. What does this tell you about the production and secretion of acetylcholine?

The synapse and drugs

Some drugs mimic acetylcholine and excite neurones. Such drugs are called stimulants; examples are nicotine, caffeine and amphetamines. Depressants have the opposite effect, they either prevent the secretion of ACH or lower its effectiveness when it reaches the next neurone. β-blockers are one example of this type of drug. They are used in the treatment of high blood pressure.

Poisons often affect the synapse. Native South American Indians use a naturally occurring poison called curare, into which they dip the tips of their hunting arrows. Curare competes with ACH for receptor sites on muscle membranes. ACH is released by the motor neurones but it cannot bind to muscle receptors, rendering the muscles useless. Curare and similar drugs can be used during surgical procedures to reduce muscular contraction. The drug LSD (lysergic acid diethylamide) affects the neurones of the brain by impairing the function of a chemical vital for normal brain function. Minute quantities of the drug can cause mental disturbances and hallucinations, which can often return unexpectedly years after the drug was first taken as 'flashbacks'.

The endocrine system

The nervous system provides a very fast and effective means of communication in the body. However, there is a second communication system – the endocrine system, which comprises specialised **endocrine glands** or groups of specialised endocrine cells found within organs. Fig 5.8 shows the position of the major endocrine glands in the human body. Endocrine glands produce hormones, which are released directly into the blood that flows through the glands. For this reason, endocrine glands are called ductless glands. Fig 5.9 distinguishes between a gland with ducts (the pancreas) and a ductless gland (the islets of Langerhans found inside the pancreas). The islets of Langerhans secrete their hormones (insulin or glucagon) directly into the capillaries that surround them. Therefore, their

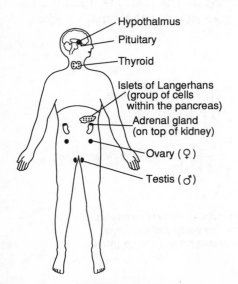

5.8　The endocrine system.

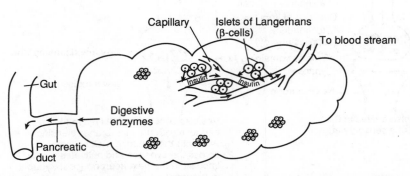

5.9　The pancreas has an endocrine gland (ductless) and a digestive gland with a duct.

products are released directly into the blood circulating around the body and they are ductless. The other cells of the pancreas produce digestive enzymes. The enzymes produced are not released into the bloodstream, but are taken into the duodenum of the small intestine by the pancreatic duct.

The function of hormones

Table 5.3 summarises the body's major endocrine glands, the hormones they produce and their effects on the body.

Table 5.3   The major endocrine glands, the hormones they produce and their effects on the body

Endocrine	Hormone	Effect
Pituitary	Thyroid stimulating hormone (TSH)	Stimulates the thyroid to produce its hormone, thyroxine
	Follicle stimulating hormone (FSH)	Initiates changes in the menstrual cycle (in the ovary) Initiates sperm formation of the testes
	Luteinizing hormone (LH)	Causes ovulation (release of ovum from the ovary) Stimulates the secretion of testosterone from the cells of the testes
	Growth hormone (GH)	Stimulates growth of skeleton and muscles
	Anti-diuretic hormone (ADH)	Reduces the amount of water lost from the kidneys in urine formation
Thyroid	Thyroxine	Controls the metabolic rate
Adrenal gland	Adrenaline noradrenaline	Prepares the body to respond to stressful situations
Pancreas (Islets of Langerhans)	Insulin (β-cells)	Lowers blood sugar levels by encouraging the conversion of soluble glucose to stored glycogen
	Glucagon (α-cells)	Raises blood sugar levels by encouraging the conversion of glucagon to glucose
Testis	Testosterone	Controls the development of male secondary sex characteristics
Ovary	Oestrogen	Controls the development of female secondary sex characteristics
	Progesterone	Prevents ovulation – hormone of pregnancy
Placenta	Chorionic gonadotrophin	Maintains the presence of the corpus luteum in the ovary, which produces progesterone

Exercise

8 Read the passage on infertility and answer the questions below.

Many couples wanting children are faced with disappointment due to infertility. There are many causes of infertility but IVF (in vitro fertilisation) has been used successfully to overcome some of them, providing many childless couples with a family. The programme usually involves giving the woman a fertility drug containing a hormone which stimulates the production of ova in the ovary. Several of these ova are removed using a long, fine needle. They are then placed in a dish containing a finely-balanced liquid medium. Conditions in the dish are carefully controlled so that the ova are not altered or damaged in any way. Semen donated from the father is centrifuged to produce a concentrated source of sperm. The sperm are then added to the ova in the dish and left, in the hope that some of the ova will be successfully fertilised by the sperm. After a couple of days, the embryos are then placed in the mother's uterus. It is hoped that at least one of the embryos will become implanted and that a normal pregnancy will follow.

a What does 'in vitro' mean?
b The fertility drug would need to contain a hormone which would increase the production of ova. Can you name such a hormone? (**Hint**: see the table of hormones and their effects on the body.)
c The ova are placed in a dish containing 'a finely-balanced liquid medium'. List the substances you think would be vital to add to the medium to keep the ova alive and healthy.
d 'Conditions in the dish are carefully controlled ...' What conditions would have to be controlled?
e How would centrifugation give rise to a concentrated source of sperm?

The pituitary gland

When we examined the nervous system, we saw that the organ in overall control was the brain. The pituitary gland (in the brain) is what controls the activities of most endocrine glands. It is sometimes called the master gland and it works in conjunction with the hypothalamus, also in the brain. The hypothalamus passes on information from the nervous system and it monitors changes in the composition of the blood. The pituitary gland secretes a number of hormones which in turn stimulate the endocrine glands to produce their own particular hormone. Fig 5.10 describes one example of how a pituitary hormone influences another endocrine gland, the thyroid.

The control of hormone release

Fig 5.10 shows that the secretion of TSH by the pituitary stimulates the thyroid to secrete thyroxine. The effect of thyroxine is to raise the metabolic rate. If the thyroid continues to secrete thyroxine, the metabolic rate will continue to rise. A system called **feedback** regulates the secretion of hormones. When the amount of thyroxine in the blood reaches a certain level, it inhibits the pituitary gland, preventing it from secreting any more thyroid stimulating hormone. This type of feedback is called negative feedback. Fig 5.11 outlines the negative feedback mechanism which maintains the composition of the blood by secreting and inhibiting the secretion of hormones.

Controlling the levels of hormones in circulation is an example of **homeostasis**. If this model is taken and applied to the secretion of two

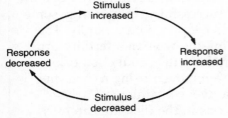

5.11   Negative feedback.

5.10   The effect of TSH from the pituitary on the thyroid gland.

other hormones, insulin and glucagon, by the islets of Langerhans in the pancreas, we will see how the level of blood sugar is kept constant – Fig 5.12.

Diabetes

Sufferers from the disease *diabetes mellitus* (usually called diabetes) produce insufficient amounts of the hormone insulin which causes higher than normal levels of glucose in their blood. The symptoms include excessive thirst, the production of sugary urine, hunger and blurred vision. Patients can go into a coma. The disease can be controlled in some cases by a carefully controlled diet, or by injections of insulin. Most diabetics can lead a normal life; indeed, former England soccer player Gary Mabbutt is a diabetic.

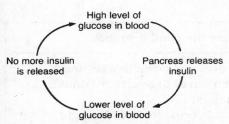

5.12   The control of glucose level in the blood

Another example of negative feedback is the regulation of the water potential of the blood (see Chapter 4). We lose a certain amount of water through the skin (sweating) and the lungs (exhaling). Most water is excreted through the kidneys in urine. If the water potential of the blood is too high (i.e. the blood is too dilute), our kidneys produce large amounts of watery urine, to rid the blood of excess water. On the other hand, if the water potential of the blood is too low (i.e. the blood is too concentrated), small amounts of concentrated urine are produced to prevent the blood becoming even more concentrated. The loss of water from the kidneys is controlled by a hormone secreted by the pituitary gland – anti-diuretic hormone (ADH). If the blood is too concentrated, ADH is secreted and prevents the kidneys from excreting too much water. If the blood becomes too dilute, the production of ADH is inhibited and more water is excreted by the kidneys.

Exercise

9   Draw a flowchart, similar to the one for the control of blood sugar level, to show the control of the water potential of the blood.

10   Drinking alcohol inhibits the production of ADH. How will this effect the concentration of the blood, if several units of alcohol are consumed?

Because the nervous and endocrine system are both involved in coordination, it is inevitable that they often operate together. At the same time, the two systems operate separately and independently, having many differences. Table 5.4 gives a comparison of the two systems.

Table 5.4   Comparison of the nervous and endocrine systems

Nervous system	Endocrine system
Communication by nerve impulses	Communication by hormones
Transmitted by neurones	Transmitted by the blood stream
Effector (muscle or gland) is stimulated	Target organ (organs) stimulated
Speed of transmission is very rapid	Speed of transmission is relatively slow
Effects are localised	Effects are widespread
Response is rapid	Response is slow
Duration of effect is short-lived	Duration of effect is long-lasting

Control and coordination in plants

Unlike animals, plants do not have a nervous system to coordinate their activities; yet they are wonderfully coordinated organisms. The ability of seeds to germinate and flowers to bloom at the right time of year and the ways in which they make the maximum use of light are two examples of the ways in which plants can respond to the environment. Plants do not possess muscles which allow them to move in response to a stimulus, but they do have the ability to grow in a particular direction in response to a stimulus such as light or water. This is known as a **tropism**.

Changes in the growth of any part of a plant are under the influence of growth substances, similar to animal hormones. Table 5.5 shows many of the tropic responses in plants.

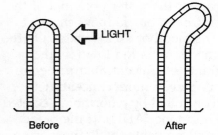

Before   After

5.13 The effect of light on a young, growing shoot.

Table 5.5   Tropic responses in plants

Tropic response	Stimulus	How the plant responds
Phototropism	Light	Leaves and some flowers angle themselves to receive maximum sunlight Shoots grow towards the light (positive phototropism) Roots (in most plants) grow away from light (negative phototropism)
Gravitropism	Gravity	Roots bend down into the soil with gravity (positive gravitropism) Most shoots grow upwards against gravity (negative gravitropism)
Thigmotropism	Touch	Some climbing plants have tendrils that twist and entwine in response to the surface they are touching, for example, the sweet pea
Hydrotropism	Water	The roots of most plants will grow towards available soil water
Chemotropism	Chemicals	Pollen tubes grow towards chemicals produced by the micropyle at the base of the ovary in a flower

How plant hormones affect growth in plants

The different ways in which plant shoots and roots respond to light (phototropism) has been very carefully studied. Read through the following experiments and try to answer the exercises as you go along.

A young growing shoot which is illuminated from one side will grow towards the light stimulus. This suggests that the growth of cells on the dark side of the shoot is greater than the growth on the illuminated side, causing the shoot to bend over. Fig 5.13 illustrates this. The dark side of the shoot differs in two ways to the illuminated side:

1 There are more cells.

2 The cells appear to have stretched or elongated.

Exercise

11 a If a growth substance or plant hormone is responsible for the difference in the lengths of the cells, how might it have affected the cell walls during cell division?
   b How do you think the elongation of the cells might have been brought about? (**Hint**: refer to water movement in cells, Chapter 4.)

We can investigate the causes of these differences in the growth of the stem by the following experiments. A very young growing stem is used, called a coleoptile. It will simply be called a shoot in these experiments.

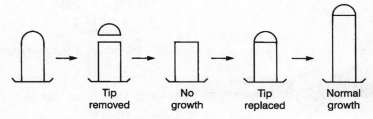

Tip removed | No growth | Tip replaced | Normal growth

5.14 The effect of the tip on the growth of a shoot.

Fig 5.14 shows that if the tip of the shoot is removed, the shoot stops growing. If the tip is replaced however, the shoot grows normally again.

Exercise

12 What does this suggest about the importance of the shoot tip in relation to the growth of the stem?

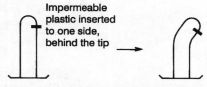

Impermeable plastic inserted to one side, behind the tip

5.15 The distribution of growth substance from the shoot tip.

Fig 5.15 shows that if a piece of impermeable material (plastic, for example) is inserted part-way into the shoot just behind the tip, an increase in growth is seen on the side without the plastic.

Exercise

13 If the growth substance is produced by the tip, how has the plastic affected its distribution?

In the experiment shown in Fig 5.16, the tips from two shoots were removed and placed on small blocks of agar (which allows substances to diffuse into it). The tips were left on the agar for different lengths of time. The agar blocks were then placed to one side of the decapitated shoots. The agar block that was in contact with a tip for the longest time produced the greatest degree of bending.

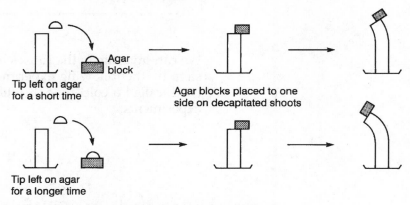

Tip left on agar for a short time

Agar block

Agar blocks placed to one side on decapitated shoots

Tip left on agar for a longer time

5.16   The effect of concentration of growth substance.

Exercise

14 a Why was the agar block able to bring about the bending in the stem?
   b Why did one block produce a greater degree of bending than the other?

The substances responsible for bringing about these growth responses have been identified as a group of chemicals called auxins – the most common auxin being indoleacetic acid (IAA).

The distribution of auxin and the way in which it affects growth in the shoot and the root is summarised in Fig 5.17. Here, a germinating seed accumulates more auxin on the lower side of the emerging root and shoot (called the radicle and plumule). The effect of this is to cause the shoot to grow bending upwards (positive phototropism) and the root to grow bending downwards (negative phototropism).

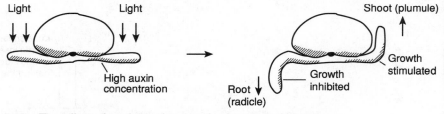

Light   Light

High auxin concentration

Root (radicle)

Growth inhibited

Shoot (plumule)

Growth stimulated

5.17   The effect of auxin on the growth of roots and shoots.

Synthetic growth substances

A number of other molecules have been identified that affect responses in plants. An understanding of these chemicals and how they function helps gardeners to get the most out of their plants. Synthetic growth substances have been developed through our knowledge of plant hormones. These include selective weedkillers. This type of weedkiller affects broad-leaved plants and disrupts their growth to such a degree that they die. It is often used on lawns to eradicate broad-leaved plants such as dandelions, the narrow-leaved grass leaves being unaffected by the chemical. Other chemicals have been developed for use as rooting powder and sprays which delay ageing in leaves, so that they have a longer shelf life in the supermarket.

15 The graph in Fig 5.18 shows how the concentration of auxin in the root and stem either stimulates or inhibits growth.

   a According to the graph, what concentration of auxin gives maximum stimulation in   i roots   ii shoots?
   b Describe the effect of auxin concentration on the growth of roots.
   c What concentration of auxin gives maximum inhibition of root growth?

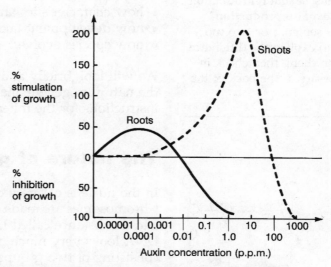

5.18   The effect of auxin concentration on root and shoot growth.

Chapter 6

Genetics and inheritance

We all expect that offspring will resemble their parents in some way. We expect that tall parents will have tall children, or that dark-haired parents will probably have dark-haired children, and so on. Genetics is the study of how this happens. It deals with:

- how characteristics are inherited
- how development takes place
- how species evolve.

We will look briefly at all of the above, but first we must understand the nature of genetic material – the material which carries the instructions for the inheritance of characteristics.

The nature of genetic material

In the nucleus of every cell is a number of chromosomes. Chromosomes are made of protein (which acts as a sort of 'scaffolding') and a substance called DNA (deoxyribonucleic acid). A strand of DNA looks very much like a ladder, with each rung in the ladder consisting of two groups of molecules. These molecules are called organic bases and are called purine bases and pyrimidine bases. Both groups are made up of two types of molecule. The purine bases are called adenine (A) and guanine (G), the pyrimidine bases are called thymine (T) and cytosine (C). If you look at Fig 6.1, you will see that the bases pair off to form the rungs of the DNA 'ladder'. Adenine will only pair with thymine and cytosine will only pair with guanine. The 'legs' of the ladder are made of two other types of molecule: a sugar (deoxyribose) and a phosphate. Fig 6.1 shows a tiny portion of the chromosome. In order that chromosomes fit into the tiny nucleus, the DNA must coil itself very tightly to reduce its volume. This is where the scaffolding protein comes in. Fig 6.2 shows how the DNA winds itself around the protein molecules to produce the compact chromosome.

Key

'Legs'
- ○ Phosphate
- ⬡ Deoxyribose

'Rungs'
- Adenine (A)
- Thymine (T)
- Guanine (G)
- Cytosine (C)

6.1 The pairing bases in a molecule of DNA.

6.2 A small section of a condensed chromosome magnified to show DNA coiling itself around protein molecules to reduce its volume.

Inside the nucleus

This chapter will be concerned mainly with DNA. This is the material that carries the instructions for all the substances a cell has to make. It is found in the nucleus as tightly coiled threads, the chromosomes. Each cell of an individual has exactly the same amount and type of DNA. In the nucleus of every human cell, there are 46 chromosomes.

Different cells may use different instructions on the chromosome to make substances. For example, every cell's DNA is capable of making the hormone insulin. However, only cells found in part of the pancreas actually use the instructions for insulin formation; in every other cell of the body, these instructions are ignored.

Cells involved in reproduction, called gametes, contain half the amount of DNA found in ordinary (somatic) cells. In humans, for example, somatic cells such as skin cells, contain 46 chromosomes while sperm and ova (eggs) contain only 23. When a sperm cell and an ovum fuse during fertilisation, the full complement of 46 chromosomes is restored.

The faithful copier

All cells have to replicate at some time. Some divide so that the organism can grow, others divide to replace worn-out cells. Skin cells, for example, are constantly being rubbed off the body and need to be replaced. When a normal body cell divides, it must make an exact copy of itself and it must especially copy its DNA without mistakes. Human somatic cells have 46 chromosomes and are said to be **diploid** (2n). **Mitosis** is the type of cell division which produces an exact copy of the original, or parent, cell. The DNA is copied faithfully, so that the new cells (daughter cells) each receive 46 chromosomes.

A second type of cell division also exists, called **meiosis**. Unlike mitosis, it results in daughter cells which have only half the number of chromosomes found in the parent cell. These cells are described as **haploid** (n). Meiosis takes place in the reproductive tissue of higher animals and plants and the haploid cells produced are called gametes. In humans, the gametes are the sperm and the ovum. Each gamete formed by meiosis is slightly different in its genetic composition from every other gamete. This is because during meiosis chromosomes often swap information, to produce daughter cells that are all genetically different. This is why you are totally unique – unless you happen to have an identical twin. Fig 6.3 summarises the two types of cell division.

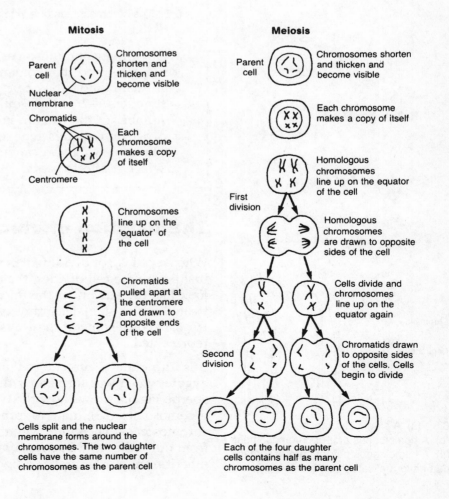

6.3   Mitosis and meiosis.

Exercise

1 Classify the following types of cell as haploid (*n*) or diploid (2*n*).

a liver cell b ovum c white blood cell d sperm cell
e muscle cell

2 Fig 6.4 shows the chromosomes of a normal woman. You will notice that the cell is about to divide, because the chromosomes have made copies of themselves. Homologous pairs of chromosomes have been placed side by side.

XX XX XX XX XX XX XX XX XX XX XX XX XX
1 2 3 4 5 6 7 8 9 10 11 12

XX XX XX XX XX XX XX XX XX XX XX
13 14 15 16 17 18 19 20 21 22 23

6.4 The chromosomes of a normal woman.

Fig 6.5 shows the chromosomes of a woman with a genetic disorder.

XX XX XX XX XX XX XX XX XX XX XX XX
1 2 3 4 5 6 7 8 9 10 11 12

XX XX XX XX XX XX XX XX XXX XX XX
13 14 15 16 17 18 19 20 21 22 23

6.5 The chromosomes of a woman with a genetic disorder.

a How many chromosomes are found in the cells of the woman in Fig 6.4 and the woman in Fig 6.5?

b If each homologous pair of chromosomes is identified by the numbers 1–23, how would you describe the chromosome number of the woman in Fig 6.5?

c A person with this genetic abnormality has Down's Syndrome. Find out what you can about the syndrome and the way in which it affects people.

The function of chromosomes

When a cell is not dividing it is very difficult to see the chromosomes. Just before the cell divides, the chromosomes appear to condense and look like threads under the microscope when the nucleus has been treated with a stain. Diagrams of chromosomes can often be confusing. Fig 6.6 shows some of the ways in which chromosomes are often represented.

It is important to remember that we all began life as a single cell, the **zygote** which was formed by the fertilisation of the ovum by the sperm. Fig 6.7 shows how the two gametes, each with a haploid set of chromosomes (23), unite to form a diploid zygote with 46 chromosomes. Therefore, we inherit two sets of chromosomes, one from our father and one from our mother. Each chromosome we inherit from our father has a 'twin' which we inherit from our mother.

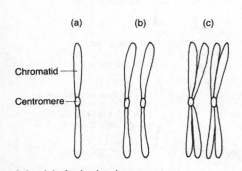

(a)   (b)   (c)

Chromatid ——

Centromere ——

6.6 (a) A single chromosome.
(b) A homologous pair of chromosomes.
(c) A homologous pair of chromosomes that have replicated during cell division.

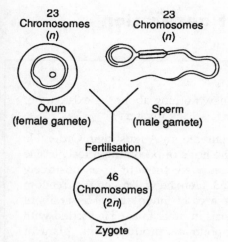

6.7 Fertilisation.

These twin chromosomes could be compared to two editions of the same book. Each book contains the same chapters containing the same information, but the information may be presented in a slightly different way. For example, chapter one might give instructions on how to make eye colour. Chapter one on the chromosome inherited from the father might give instructions for making blue eyes, while the same chapter on the twin chromosome inherited from the mother might give instructions on how to make brown eyes. These pairs of chromosomes are called homologous chromosomes (*homo* = the same).

Each chromosome of a homologous pair carries information, or chapters, in the same place. Chapter two always follows chapter one. Each chapter is called a **gene** and its position on the chromosome is called its **locus**. Each gene is a set of instructions for making a particular protein, for example, the hormone insulin. It is vital that the instructions contained in a gene are correct. If any of them are incorrect or missing, then the results could be disastrous for the organism.

Dominant and recessive traits

The inheritance of eye colour in humans is thought to be much more complex than this, but it is simplified as much as possible to help your understanding.

The genes found at the same position or locus on a homologous pair of chromosomes are called **alleles**. Alleles of a gene can be the same, or they can differ slightly from one another. We could look at the example of eye colour mentioned in the last section. Imagine that the inheritance of eye colour is controlled by one gene with two alleles, one allele instructs for brown eyes and the other instructs for blue eyes. If we call the allele for brown eyes *B* and the allele for blue eyes *b*, it is possible to inherit three different combinations of these alleles, depending on which is contained in the sperm or ovum. Fig 6.8 shows how the allele combinations are inherited and that the three possible combinations are *BB*, *Bb* or *bb*. If the combination was *BB* the child would have brown eyes. If it was *Bb*, the child would also have brown eyes. Only the *bb* combination would produce blue eyes. In the *Bb* child, the allele for blue eyes has been hidden, or masked. This is because the allele for brown eyes (*B*) is said to be **dominant**, while the allele for blue eyes (*b*) is **recessive**.

The type of genes that a person carries on his or her chromosomes is called the person's **genotype**. How these genes affect us or how they make us look is called our **phenotype**. If the alleles inherited are the same, i.e. *BB* or *bb*, we say that the genotype of this person is **homozygous** and if different alleles are inherited, i.e. *Bb*, then the genotype is **heterozygous**.

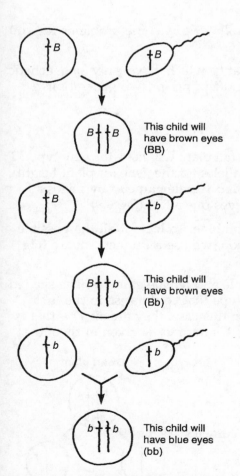

6.8 Possible combinations of alleles for blue and brown eyes.

3 In some plants, it is possible to have either red flowers or white flowers. The colour is controlled by a gene having two alleles.

  a Give the alleles (red and white) two letters to represent them. The letters should describe the gene for flower colour as having a dominant red allele and a recessive white allele.

  b Which combination(s) of alleles will give rise to i white flowers ii red flowers?

4 Write brief explanations of the following terms and then check them with the glossary: gene, allele, locus, genotype, phenotype, recessive, dominant.

Handing on to the next generation

The work of Mendel

How are our genes or characteristics handed down to our offspring? If you knew nothing about the existence of chromosomes or genes, or of how they worked, how would you begin to unravel the mystery?

In the nineteenth century an Austrian, Gregor Mendel, became a priest and entered an Augustinian Order. He studied physics, maths, zoology and botany at the University of Vienna in the hope of becoming a teacher. He quickly found that independent thought was not encouraged and new ideas were thought of as positively outrageous! Mendel failed his exams and entered a monastery in Brunn in 1853. Here he was given the freedom to indulge his love for maths and botany and his fascination with the whole area of inheritance. He set about breeding pea plants and statistically monitoring the results, to establish patterns in inheritance. He started with 'true-breeding' plants, i.e. giant or dwarf plants that, when allowed to self-pollinate, produced only giant or dwarf plants generation after generation.

He would then cross-pollinate these contrasting plants and analyse the offspring. Mendel identified that each individual possessed two 'factors' which determine a specific characteristic. He also showed that a parent only transmitted one of these factors to any offspring.

In 1865, Mendel presented his results to the Natural History Society of Brunn and in the following year they were published in a journal. It is sad that Mendel was not taken seriously by his learned colleagues of the day. His work remained largely unknown until 1900 and in 1909 the term 'gene' was coined by the Danish botanist Johannson.

Working with true-breeding plants that have only one difference to monitor (a **monohybrid cross**), we can predict the genotype of the offspring.

Here are some worked examples, followed by some problems for you to work out for yourself.

If we cross two true-breeding plants which have only one genetic difference, how can we predict the offspring for two generations?

Look at the following cross:

Tall plant × Dwarf plant

Because they are true-breeding, the tall plant will have the genotype TT and the dwarf plant tt (T and t are alleles for the gene for plant height). During the formation of the male and female gametes by meiosis (gametogenesis), each gamete receives one allele (Fig 6.9).

During fertilisation, the gametes combine. To look at all the possible combinations and to avoid confusion, we use a Punnett square (Fig 6.10).

In the white squares we put the male and female gametes. The shaded squares show every possible genotype that could arise in the first generation, or the F_1 **generation**. In this case, they are all Tt – that is they are all heterozygous and tall. A summary is given in the table.

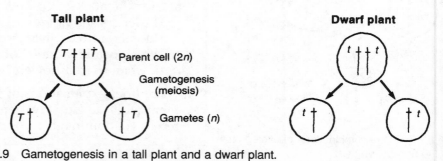

6.9 Gametogenesis in a tall plant and a dwarf plant.

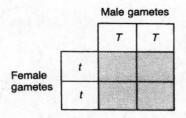 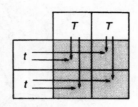

Male gametes

Female gametes

6.10 How Punnett squares are used to predict the offspring of a monohybrid cross.

Gametogenesis

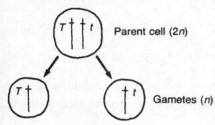

Parent cell (2n)

Gametes (n)

Genotype	Phenotype
100% heterozygous (*Tt*)	tall

To work out the genotype and phenotype of the offspring in the second or **F$_2$ generation**, we take two of the offspring from the F$_1$ generation and look at the gametes they produce. Then, using the Punnett square, we look at all the combinations of these gametes (Fig 6.11).

Punnett square

	T	*t*
T	*TT*	*Tt*
t	*Tt*	*tt*

} F$_2$ generation

6.11 Gametogenesis in the F$_2$ generation and a Punnett square showing the genotypes of the F$_2$ generation.

Genotype	Phenotype
50% heterozygous (*Tt*)	tall
25% homozygous dominant (*TT*)	tall
25% homozygous recessive (*tt*)	dwarf

This tells us that the genotypes of the second generation will be in the ratio 1(*TT*):2(*Tt*):1(*tt*) and that the phenotypes will be in the ratio of 3(tall):1(dwarf).

Plants with the genotype *TT* or *Tt* will both be tall. How is it possible to work out which of the genotypes it possesses since you cannot tell which alleles it has inherited simply by looking at it? The best way is to perform a **test cross**. The plant with the dominant characteristic (tall) is crossed with a plant showing the recessive characteristic (dwarf). Fig 6.12 shows the result of such a cross. If the tall plant was homozygous (*TT*), then all the offspring will be tall (*Tt*). If the tall plant was heterozygous (had a hidden recessive gene) then a 1:1 ratio will result, in other words half the offspring will be tall and half will be dwarf.

	T	*T*
t	*Tt*	*Tt*
t	*Tt*	*Tt*

6.12 A test cross determining that the tall plant was homozygous.

We will look at another example for monohybrid inheritance. This shows how the ability to roll the tongue can be inherited in humans. Again this is oversimplified, it is believed that the inheritance of this ability is much more complex than the example shows. In fact there are few 'simple' examples of monohybrid inheritance in human genetics.

Example

If one gene controlled the ability to roll the tongue and it had two alleles, *R* (dominant) and *r* (recessive), a person with the tongue-rolling ability would have the genotype *RR* or *Rr*. Those with *rr* genotypes would be unable to roll their tongues. If we begin with homozygous parents (*RR* × *rr*) and follow the steps in the monohybrid cross, it is possible to predict every possible genotype for their children (Fig 6.13).

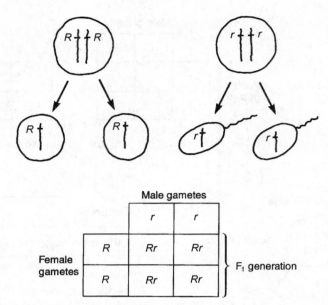

6.13   A Punnett square showing the genotypes in the F₁ generation (tongue rolling).

Genotype	Phenotype
100% heterozygous (*Rr*)	able to roll tongue

If one of these individuals had children with another heterozygous individual (*Rr* × *Rr*), the results would be as shown in Fig 6.14.

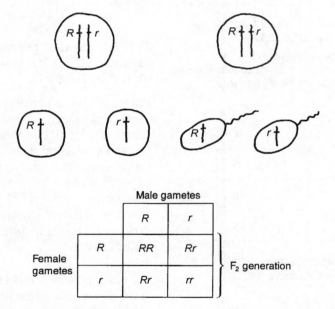

6.14   A Punnett square showing the genotypes in the F₂ generation (tongue rolling).

Genotype	Phenotype
50% heterozygous (*Rr*)	able to roll tongue
25% homozygous dominant (*RR*)	able to roll tongue
25% homozygous recessive (*rr*)	unable to roll tongue

Here we see the same pattern of ratios in the F₂ generation as we saw in the previous example – a 3 : 1 ratio.

Exercise

5 An albino mouse has the genotype cc and a mouse with a normal brown coat has the genotype *CC*. Use Punnett squares to work out the genotypes and the phenotype ratios of the F_1 and F_2 generations.

6 Certain varieties of tomato plant have hairy stems while others are hairless. The gene for hairy stems (*H*) is dominant to the gene for hairless stems (*h*).

   a If a tomato plant has hairy stems, what are its two possible genotypes?
   b A hairless plant can only have one genotype. Why?
   c A hairy-stemmed plant is crossed with a hairless one and the following phenotypic ratio is obtained: 50% hairy stems and 50% hairless stems. Explain this result.

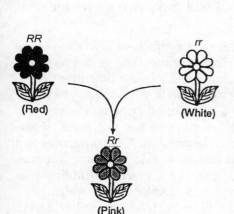

6.15 Codominance in plants.

Do dominant genes always dominate?

The cases examined so far have only ever resulted in two different phenotypes. This is because one of the alleles for the gene has always been able to dominate. If we examine the following cross, we will see that there can be exceptions.

Look carefully at the cross in Fig 6.15. Can you work out why the offspring is not red? If the gametes are placed in a Punnett square, we will get the expected genotype (*Rr*) in all the offspring, but the phenotype is a colour somewhere in between the phenotype of the parents. This mixing of red and white produces a pink colour. In this example, both alleles of the gene for flower colour appear to show equal strength or dominance. This is known as **codominance**.

Exercise

7 Using the worked examples of monohybrid crosses and taking alleles to be codominant, predict the genotypes and phenotypic ratios of the following crosses.

   a blood group A (genotype *AA*) with blood group AB (genotype *AB*)
   b blood group B (genotype *BB*) with blood group AB
   c blood group AB with blood group AB

How sex is determined

Whether you are male or female is determined by a special pair of chromosomes called the sex chromosomes. These chromosomes are also known as the X and Y chromosomes. Females inherit two X chromosomes and males inherit an X chromosome and a Y chromosome. In the ovaries of the female, gametogenesis (meiosis) results in every ovum receiving one X chromosome. In the testes of males, half of the sperm cells receive an X chromosome and half receive a Y chromosome.

The sex of a baby is determined by the type of sex chromosome contained within the sperm. If a Y-bearing sperm fertilises an ovum, then the baby will be a boy. If an X-bearing sperm fertilises the ovum, then the baby will be a girl (Fig 6.16).

8 King Henry VIII executed one wife and divorced another because they did not give him a son. Can you give an explanation for why no boys were born?

Sex-linked inheritance

In humans, the X chromosome is larger than the Y chromosome. Genes located at the top of the X chromosome are therefore absent from the Y chromosome (Fig 6.17). Each of the other 22 homologous pairs of chromosomes are of equal length and will carry two alleles for each gene.

The gene for the production of blood-clotting factors is known to be present on the upper part of the X chromosome. The gene has two alleles, the allele for normal blood clotting (H) being dominant to the allele for haemophilia (h). A normal female will possess two dominant alleles (HH). However, if a woman has one recessive (h) allele, this will be transmitted to half of the ova she produces. If an h-bearing ovum is fertilised by a sperm bearing an X chromosome with a normal H allele, the daughter that results would not have haemophilia, but is said to be a carrier. If, however, an ovum containing a recessive (h) allele on its X chromosome is fertilised by a sperm containing a Y chromosome, the son will be born haemophiliac, having no allele on the Y chromosome to mask the effects of the h allele.

Because the blood-clotting gene is found only on the X chromosome, it is important to show whether the person is male or female and which alleles they possess. For this reason, we identity the chromosome as well as the allele – this shows the genotype, the sex and the phenotype.

Fig 6.18 shows gametogenesis in a female carrier and in a normal male. The Punnett square shows the possible combinations of the alleles in the offspring.

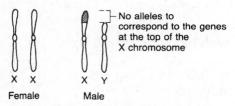

6.16   Inheritance of sex in humans.
(a) Gametogenesis in the male and female.
(b) A Punnett square showing the possible combinations after fertilisation.

6.17   Structure of the X and Y chromosomes in humans.

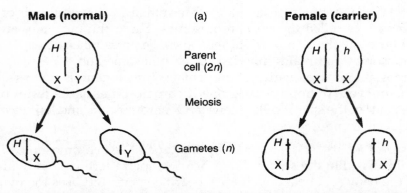

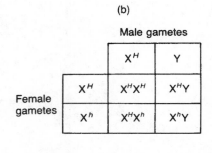

6.18   Sex-linked inheritance of the blood clotting factor gene. (a) Gametogenesis in the male and female. (b) A Punnett square showing the possible combinations of alleles after fertilisation.

These results are summarised in the table below.

Genotype	Sex	Phenotype
25% $X^H X^H$	Female	Normal
25% $X^H X^h$	Female	Carrier
25% $X^H Y$	Male	Normal
25% $X^h Y$	Male	Haemophiliac

Exercise

9 Why is it rare to find a female haemophiliac?

10 When a woman carrying two normal alleles on her X chromosomes marries a haemophiliac man, what are the possible genotypes and phenotypes of their offspring?

11 The royal families of Europe have a history of haemophilia. Fig 6.19 shows part of the pedigree of Queen Victoria. Look at the children of Beatrice and Henry (whose genotypes have not been shown). Use the genotypes of the children to predict the genotypes and phenotypes of Beatrice and Henry.

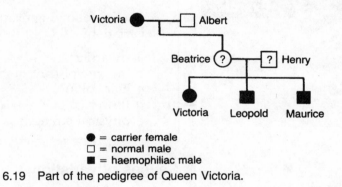

● = carrier female
□ = normal male
■ = haemophiliac male

6.19 Part of the pedigree of Queen Victoria.

12 What possible genotypes could the parents of a female haemophiliac have?

13 In humans, normal colour vision is a sex-linked trait. The gene which controls normal vision has two alleles, C for normal vision is dominant to c for red–green colour-blindness. If a colour-blind woman marries a man with normal vision, work out the possible genotypes and phenotypes of their offspring.

Variation and selection

Within any species, there is a tremendous amount of variation. People are all different. They are of different heights, they have different coloured skins, different coloured eyes and different shaped ears and noses. The variation between individuals in a group can be **continuous** or it can be **discontinuous**. An example of continuous variation in humans is height. It shows a gradual transition between two extremes – very small to very tall. In a group of people you will find a few that are very small, a lot of medium-sized people and a few tall

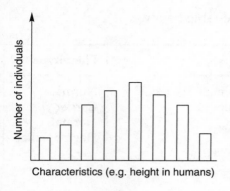

6.20   Continuous variation produces a normal distribution.

A normal distribution curve is bell shaped with most of the population around the mean value and fewer around the two extremities.

people. If the heights of a large number of people are measured and the results plotted as a histogram, we get a graph which shows a normal distribution curve. Fig 6.20 shows the normal distribution curve for height in a population, showing the smooth transition from one height to another without a break. Continuous variation can be caused by the combined effects of several genes. The isolated effect of any one of the genes is minimal, but all the genes combined, together with environmental influences, have a considerable effect. If a plant's genotype is *TT*, for tallness, this height will only be attained if the plant receives sufficient light, water and the right temperature – the limiting factors. If environmental conditions are unfavourable, the gene will be prevented from expressing its full effects.

Sometimes there are no 'in-betweens'. For example, you can either roll your tongue or you cannot. You are either male or female, your blood will belong to one of the blood groups A, B, AB or O. This type of variation is called discontinuous variation.

Fig 6.21 shows the frequency distribution in a case of discontinuous variation.

Discontinuous variation is caused by one or two genes which may have two or more alleles. The environment has little or no effect on the way in which the genes express themselves.

14 Do the following examples show continuous or discontinuous variation?

    a dwarfism
    b the size of fruit on a pear tree
    c hair colour
    d flower colour
    e physical strength

Causes of variation

Inherited variation is thought to be brought about in the following ways:

- Genetic information is swapped between chromosomes of a homologous pair during meiosis. Fig 6.22 shows how sections of a chromosome randomly cross-over. The result is that each chromosome involved may possess a new combination of alleles for the genes on that section of chromosome.
- When chromosomes align themselves on the spindle in the first phase of meiosis, the way in which they do this is completely random. Fig 6.23 illustrates this independent assortment.
- In sexual reproduction, the pooling of genetic information from two gametes will give rise to unique offspring.

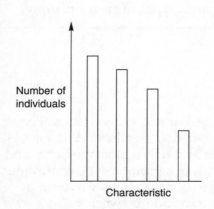

6.21   Frequency distribution for discontinuous variation.

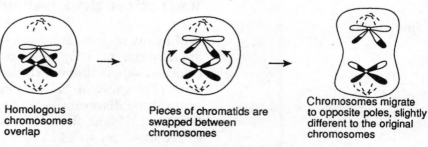

Homologous chromosomes overlap

Pieces of chromatids are swapped between chromosomes

Chromosomes migrate to opposite poles, slightly different to the original chromosomes

6.22   How crossing-over during meiosis leads to variation.

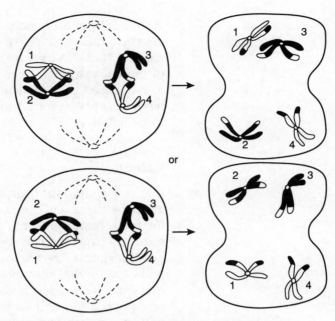

or

6.23 Independent assortment of chromosomes during meiosis.

- Genes often undergo changes or mutations. These changes alter the type of protein they produce.

Mutations

A mutation is a spontaneous change to a chromosome or a gene that causes it to produce a different characteristic. Mutations can arise because of a mistake during cell division, when the DNA is not copied exactly. It can also be caused by agents called **mutagens**. Examples of mutagens are radiation and certain chemicals like those found in cigarette smoke.

Mutations can occur in ordinary (somatic) cells or in gametes. If they occur in somatic cells, they usually go undetected. If they happen in the gametes, this is usually detrimental for the offspring. Mutations are seldom beneficial. A mutation can be a change to a single gene, to a number of genes or even to the number of chromosomes.

Mutations that are not beneficial

Down's syndrome
During meiosis in the formation of gametes, one gamete ends up with two number 21 chromosomes. It follows that another gamete ends up with no number 21 chromosome. If the gamete without a chromosome 21 is fertilised, then the offspring does not survive. If the gamete with two chromosome 21s is fertilised, then the offspring ends up with three chromosome 21s and has Down's syndrome. People with Down's syndrome are usually mentally retarded and have characteristic facial features.

Kleinfelter's syndrome
This results from inheriting an extra X chromosome in men. The affected men are sterile and usually mentally retarded.

Turner's syndrome
This occurs in women who inherit only one X chromosome. The affected female is sterile, does not menstruate and is short in stature. Often, she may have a web of skin on either side of the neck, abnormalities of the aorta and elbows that angle outwards at the elbow.

About 15% of all conceptions terminate in either a spontaneous abortion (miscarriage) or stillbirth. There is growing evidence to suggest that this may be a result of abnormalities in the chromosomes of the baby. In fact, over 50% of early miscarriages have been shown to be caused by genetic abnormalities. Some of these abnormalities have never been found in live-born babies and are therefore presumed to be lethal.

Selection

In any population, from tiny microorganisms to large plants and animals, there will always be competition. Animals have to compete for food, homes and mates. Plants compete for light, water and minerals from the soil. When resources are scarce, only the fittest, healthiest, most fertile and well-adapted members of the population will obtain what they need to survive from the environment. These members have what is called 'survival value'. The weaker members of a population will be less likely to mate. Those with survival value will live longer and pass on their genes to more offspring. This is how natural selection works. The pressures of the environment force the weaker members of the population to die out and their genes die with them.

Exercise

15 Think of examples of another type of selection, artificial selection, where animals are deliberately bred because of their appearance, bringing about changes in the species that are not always beneficial to the animal.

Evolution

The notion of a theory to explain evolution did not begin with Darwin when he published his theory on evolution. Long before Darwin published, people were attempting to explain diversity. In 1809, the French biologist Jean-Baptiste Lamark put forward a hypothesis to explain evolution. He proposed that change was a result of minor changes that an individual acquired in a lifetime, through use or misuse of body parts. The long neck of the giraffe was a result of the animal stretching to reach the leaves on tall trees. This stretching would cause it to develop a slightly longer neck and this increase in length would be passed on to its offspring. The neck would be further stretched by the offspring.

In a similar way, he proposed that misuse of certain limbs or organs would lead to them becoming smaller and disappearing from the species.

The amazing diversity of species found by Charles Darwin on his voyages to the Galapagos Islands, led him to begin to think about how these species arose. He was also fascinated by the amount of diversity within a species. On his travels as a naturalist aboard the survey ship *H.M.S. Beagle*, he collected an enormous amount of information concerning the amount of variation between organisms, which convinced him that species were capable of changing. He understood that organisms had the potential to produce a large number of offspring, yet populations appeared to remain fairly constant. By 1839 he had put together a theory of evolution, based on natural selection, which attempted to explain how evolution might occur. Darwin knew that his theory would have a profound impact on the way people thought and was very reluctant to publish his theories.

At the same time, Alfred Russel Wallace, another naturalist who had travelled extensively, had come to the same conclusions as Darwin. He wrote a twenty-page essay which outlined his theory of evolution and in 1858, he sent the essay to Darwin. In 1859, Darwin published *On the Origin of Species by Means of Natural Selection.*

It is alleged that the work was almost complete when he received the essay from Wallace. Because he feared the response the book would evoke, he had planned for it to be published after his death. The essay from Wallace altered his plans.

One example of evolution in action through natural selection is the peppered moth. The natural colour of the moth is light and speckled. In 1850 the light variety was the most common type. Its colouring increased its survival value, since it was very well camouflaged on the lichen-covered tree trunks. By 1895, in the industrial areas of England, a black variety of peppered moth had risen to account for 98% of the population. Industrial pollution had meant that black, dirt-covered tree trunks were now more common in industrial areas.

This type of selection occurred because of a mutation which produced the dark variety which was well adapted to the environment at that time. This is not to say that this mutation had never occurred before. However, if it had, the black variety would have been very conspicuous and easily seen by predatory birds. Mutations like this give rise to beneficial changes, allowing a species to adapt to changes in the environment and even preventing their extinction. However, if the changes in the environment or climate are too rapid, this is not always possible.

Fig 6.24   Dark and light forms of the peppered moth on a lichen-covered tree trunk.

Exercise

16 Why do you think that organisms that depend totally on asexual reproduction have difficulty in adapting to changes in the environment?

Microbes

STARTING POINTS

● You should be able to draw and label diagrams of a simple plant cell and a simple animal cell. Check with the answer section before continuing.

Introduction

Microbiology is the study of **microorganisms**. These are the members of a large group of free-living cells that can exist singly or as clusters. This is one of the basic differences between cells of microbes and those of animals and plants, which can exist only as part of a larger **multicellular** organism. A single microbial cell is able to carry out all the processes necessary for life – nutrition, respiration, excretion, reproduction and growth – independently of other cells.

Cell structures

All organisms fall into two categories, **prokaryotes** and **eukaryotes**, as shown in the table opposite.

The important difference between the two groups is in the structure of the cell nucleus. Prokaryotes have a single strand of DNA free in the **cytoplasm**, whereas in eukaryotes the DNA is enclosed in a membrane, forming a true nucleus. This suggests that eukaryotes evolved from prokaryotes.

Prokaryotes have another difference – they possess a **cell wall** (different from that of a plant cell) which gives the organism its rigidity and shape.

Types of organisms

Prokaryotes	Eukaryotes
Blue–green algae	Fungi
Bacteria	Algae
	Plants
	Animals

Bacteria

Bacterial cells are extremely small, ranging in length from 1 μm to 20 μm (1 μm = 0.001 mm). They are generally classified according to their shape. Fig 7.1 shows the shapes of different types of bacteria.

Fig 7.2 shows the structures within a generalised bacterial cell. We will see how some of these structures give bacteria survival value.

The cell wall

The bacterial cell wall is made up of a complex mixture of proteins, sugars and lipids (fats). This is unlike the plant cell wall, which is made up of cellulose. Some bacteria have a **slime capsule** outside their cell wall.

There are two types of cell walls to be found in bacteria so we can divide them into two distinct groups – **Gram-positive** bacteria, G^+, and **Gram-negative**, G^-. These names derive from the staining technique, Gram's method (used to view bacteria more easily under the microscope), which gives different results according to the type of cell wall they possess. Fig 7.3 shows the different structures of the G^+ and G^- cell walls. G^+ bacteria have a thick, single-layered wall, while G^- bacteria have two thin walls.

Sphere (coccus)		
○ ○ ○		
○○	pair	e.g. *Diplococcus*; causes pneumonia
○○○○○○	chain	e.g. *Streptococcus*; causes sore throats
🍇	clump	e.g. *Staphylococcus*; causes boils
Rod (bacillus)		
○ ○ ○ ○ ○ ○		e.g. *Escherichia coli*; harmless bacterium found in the human intestine, and *Clostridium*; causes tetanus
Bent rod (vibrio)		
؏		e.g. *Vibrio cholerae*; causes cholera
Spiral (spirillium)		
～～～		e.g. *Treponema*; causes syphilis

7.1   Classification of bacteria according to shape.

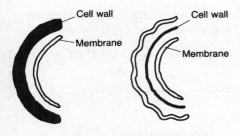

7.3   Cell walls of Gram-positive and Gram-negative bacteria.

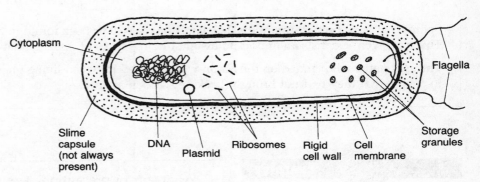

7.2   General structure of a bacterium.

The cell membrane

Sometimes called the plasma membrane, this is a thin structure that completely surrounds the cell. This vital structure is a critical barrier separating the inside of the cell from its environment. If the membrane is broken, the cell will die.

Exercise

> 1 If you had to design an agent which would kill bacteria (a **bactericide**), how would you make use of the last piece of information given above?

Inside the cell

The DNA in bacterial cells is a highly folded, single chromosome which forms a loop or circle. In many bacteria, a separate circle of DNA has also been discovered. This is called a **plasmid**. A plasmid is completely independent of the chromosome. It can make copies of itself and, most importantly, it can transfer copies of itself to other bacteria. We will see why this is important later.

Also found in the cytoplasm are **storage granules**, containing

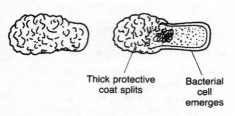

7.4 Bacterial spores.

glycogen, lipids and other food reserves, and **ribosomes**, which are responsible for protein synthesis.

Some bacteria possess tail-like structures called **flagella**. These bring about movement by a whip-like action, propelling the cell along.

How bacteria survive harsh conditions

A characteristic of certain rod-shaped bacteria is that under certain conditions, usually when food supplies are exhausted or they are running short of water, they are able to form a resting cell called an **endospore**. This allows the bacterium to survive very high temperatures and changes in pH. The thick protective coat of the spore allows the cell to remain dormant for a very long time. When suitable conditions return, the spore germinates to produce a single bacterial cell (see Fig 7.4). This has very serious implications for food storage and preservation.

Spores

Spores that have been dormant for thousands of years have been found in the tombs of the Egyptian mummies! This has given scientists an important insight into the types of diseases that the ancient Egyptians suffered from.

The astronauts who landed on the Moon were isolated in sealed living quarters for some days after their return to ensure that they had not brought back any 'Moon spores'.

Exercise

2 What sort of precautions would you take against spores when storing, preserving or preparing food?

7.5 Binary fission in bacteria.

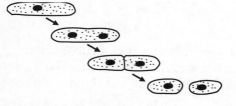

7.6 Typical growth curve for a bacterial population.

Reproduction

Bacteria reproduce mainly asexually, by a method known as **binary fission** (see Fig 7.5). The cell grows until it reaches its maximum size, the DNA replicates itself and the cell divides to give two identical daughter cells. This process takes about 20 minutes. It has been estimated that if division continued at this rate, after 26 hours a single bacterium (0.001 mm in length) would have produced enough offspring to occupy a volume of 1000 m^3 (about the size of a house)!

This, of course, would not happen in practice, since many bacteria would die before they reproduced, nutrients would run out, or they would be poisoned by their own toxic waste.

Fig 7.6 shows a typical growth curve for a population of bacteria grown on nutrient agar.

The graph is divided into four phases as follows:

A **The lag phase**. The bacteria are 'settling in', becoming adjusted to their new environment.

B **The logarithmic phase**. The bacteria begin to divide and multiply. This phase is also known as the phase of **exponential growth** – each cell in the population is dividing to form two cells.

C **The stationary phase**. This shows no increase or decrease in the numbers. This may be due to competition between bacteria for food, space or oxygen. A build-up of wastes may cause undesirable changes in the environment.

D **The death phase**. The bacteria begin to die because of the lack of oxygen and food and because of the build-up of toxic waste.

Nutrition

Like all organisms, bacteria require a constant supply of nutrients in order to obtain energy and the raw materials needed to build their cells. They do this in a wide variety of ways (see box).

Nutrition in bacteria

Autotrophic bacteria. These are capable of synthesising their food from inorganic materials. They can be subdivided into the **photosynthetic** autotrophs and the **chemosynthetic** autotrophs.

- Photosynthetic autotrophs use sunlight as a source of energy, for example, the blue–green algae.
- Chemosynthetic autotrophs get their energy from chemical reactions, for example the nitrifying bacteria in the nitrogen cycle.

Heterotrophic bacteria. These cannot synthesise their own food and require a supply of organic material. They can be subdivided into **saprophytes** and **parasites**.

- Saprophytes use non-living sources of organic material, for example the bacteria of decay.
- Parasites use a living source of organic material, for example disease-causing bacteria.

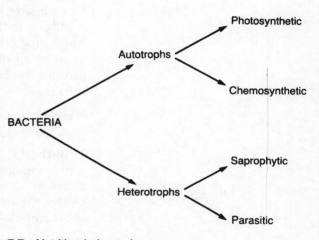

7.7 Nutrition in bacteria.

Bacterial respiration

Bacteria can respire aerobically or anaerobically. Most bacteria are aerobic. Anaerobic bacteria may produce lactic acid, acetic acid or ethanol as a by-product. Some bacteria can only respire in the absence of oxygen. Others can respire with or without oxygen.

Bacteria and humans

Some bacteria can be very harmful to us while others play an essential role in our lives.

An example of bacteria that are harmful to us is those which take nitrogen from the soil and reduce its fertility.

Exercise

3 Give some other examples of bacteria which are harmful to humans.

Pathogens

Pathogens are parasites. A parasite is an organism that lives on and causes damage to another living organism, called the **host**. The two form a host–parasite relationship and the outcome of this relationship depends on two factors:

- How pathogenic (or **virulent**) the parasite is, in other words, how much damage it can inflict on the host.
- How susceptible or resistant the host is to the parasite.

When pathogens enter the body, we become infected. This does not mean that we have a disease, but if the parasite is not checked and destroyed, its numbers may rise to a level which causes disease.

Pathogenic organisms can gain entry to the body through the skin (via a wound), or through mucus membranes in the lungs or the intestines (from our food).

Once inside the body, in order to do any damage the pathogen must multiply. Pathogens bring about damage to the host in many ways. Most produce **toxins** (poisonous chemicals) which are responsible for most or all of the damage. When released, these toxins can spread around the body, damaging tissue far away from the site of the infection.

Here are some examples of diseases caused by toxin-producing bacteria.

Diphtheria. This disease is transmitted by droplet infection, that is, when someone already infected coughs or sneezes, the droplets of saliva sprayed from the mouth carry the bacteria. The toxin produced causes the lining of the air passages to become inflamed. It destroys heart muscle and other tissues.

Tetanus. Sometimes called lockjaw, this disease develops as the result of a deep cut from a dirty object, such as a nail. The toxin attacks the nervous system and causes muscle paralysis.

Botulism. The bacterium itself rarely grows inside the body, but does grow and produce the toxin in improperly canned food. Eating the food results in paralysis and is often followed by death. This toxin is one of the most poisonous substances known. One milligram of the pure toxin is enough to kill more than one million guinea pigs. Death is usually due to respiratory or cardiac failure.

Human defences against disease

The body is constantly being invaded by pathogens. They enter through our orifices, particularly through mouth and nose. To defend against the invasion, we have an immune system. We have two types of immunity, natural immunity and acquired immunity.

Natural immunity

Our skin is our first line of defence. Bacteria can only enter if the skin is broken or cut. If this occurs, blood clots will quickly form to prevent further entry. Any bacteria that do enter through a wound will usually be destroyed by special white blood cells, called **phagocytes**. These cells engulf the bacteria and destroy them (Fig 7.8). In the process, the phagocytes themselves die, forming the yellow pus found on an infected cut.

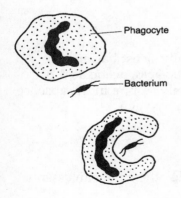

— Phagocyte

— Bacterium

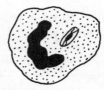

7.8   The action of a phagocyte as it surrounds, engulfs and destroys a bacterium.

Tears and mucus secretions have been found to contain an enzyme called lysozyme, which is capable of destroying bacteria even at low concentrations. Mucus secretions, together with the cilia lining the trachea and bronchi of the lungs, help to trap foreign bodies and push them back into the mouth. Most bacteria swallowed will be destroyed by the acid in the stomach. If bacteria do multiply in the stomach, we often vomit – thus removing them. Also found in our bloodstream is a system of enzymes known as **complement**. If these enzymes are activated, they will help our acquired immune system to destroy pathogens.

Acquired immunity

Acquired immunity is so called because we are not born with it; it is something we build up over a period of time. The key to acquired immunity is memory – the ability of the immune system to recognise a foreign body if it encounters it again after the first infection.

Bacterial cells have on their surfaces chemical substances called antigens. When we become infected with a pathogen for the first time, another group of white blood cells, the **B lymphocytes**, learn to recognise the antigens. The lymphocytes are produced by the bone marrow. They produce antibody proteins in response to the antigens. The antibodies then attach themselves to the bacteria. This either causes the bacteria to clump together, making it easier for the phagocytes to engulf them, or neutralises the bacterial toxin.

7.9   Production of antibodies in response to a specific type of antigen.

The antibodies produced by lymphocytes are very specific; that is, they will attack only one type of bacterial cell (see Fig 7.9). Once the body has been infected, lymphocytes learn to recognise the particular bacterium and are prepared to produce antibodies very quickly the next time it invades.

Exercise

> 4   AIDS stands for Acquired Immune Deficiency Syndrome. People with the illness do not die from AIDS, but from repeated common infections such as influenza or pneumonia. Can you explain why this is so?

How antibodies are formed

When an antigen is present in the blood stream, it is thought to bind to a special receptor site on the surface of the lymphocyte. This binding causes the lymphocyte to multiply and produce a large number of identical cells or clones. These clones are capable of producing large numbers of antibodies very efficiently, as shown in Fig 7.10. The antibodies binding to the antigens cause them to form clumps. These

7.10 The binding of antigen to B lymphocytes to produce antibody-producing clones.

clumps can then be destroyed by phagocytes or they can activate the complement enzymes already present in the blood, which destroy the bacteria. It is believed that long before we are born, we develop an enormous range of lymphocytes, each with the potential to produce an antibody against a specific antigen. Because all cells in our body have surface antigens, it is believed that before we are born, our lymphocytes learn to ignore these antigens, so our own cells will never be attacked.

Organ transplants

A greater understanding of how our immune system works has led to our being able to transfer organs to replace diseased or damaged ones. The organs introduced are sure to be thought of as 'foreign' by the immune system, because their cell surfaces possess antigens that it does not recognise. These cells will be attacked directly by another type of lymphocyte that does not produce antibodies – the T lymphocytes. They look just like B lymphocytes, but they have special molecules on their cell surfaces that stick to antigens on the surfaces of cells. They will not attack the body's own cell antigens, but will attack the antigens on the surface of the donated organ. This can result in the organ being rejected. To help prevent this from happening, the patient can be given a drug, called an immunosuppressant, to limit the production of these lymphocytes. This does, however, make it a dangerous time for these patients. If the body is invaded by pathogenic bacteria, they will have little resistance to infection.

Antibiotics

Antibiotics are chemicals produced by microorganisms that kill or prevent the growth of other microorganisms, without damaging the host.

Antibiotics

Here are some examples of common antibiotics:

Penicillin: Discovered by Alexander Fleming in 1928. It is produced by the fungus *Penicillium notatum*. Penicillin works by preventing the synthesis of the cell wall when the bacteria are dividing. This leaves the cell vulnerable to attack by phagocytes.

Fleming discovered penicillin following the accidental contamination of a bacterial culture by a mould – usually an indication of sloppy technique! The mould killed the growing bacteria. Surprisingly, Fleming did not follow up his observations and it was some years later that a team including Howard Florey, Ernst Chain and N G Heatley extracted enough penicillin to try it out on patients. The results were spectacular. During World War II, penicillin saved the lives of thousands of soldiers who would otherwise have died from infected wounds.

Tetracycline: This is effective against a wide range of bacteria. For this reason, it is called a **broad spectrum antibiotic**.

Exercise

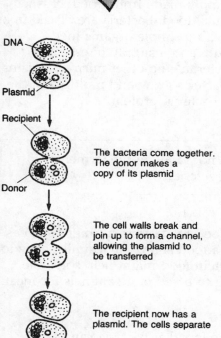

DNA

Plasmid

Recipient

The bacteria come together. The donor makes a copy of its plasmid

Donor

The cell walls break and join up to form a channel, allowing the plasmid to be transferred

The recipient now has a plasmid. The cells separate

7.11 Transfer of plasmids between bacteria.

5 Some antibiotics are taken orally (by mouth). What important features must they have in order to be taken in this way?

Antibiotic resistance

Penicillin is not nearly as effective against pathogens as it used to be. More and more bacteria have become resistant to it, as is the case with many of our popular antibiotics. This is because for many years antibiotics were over-prescribed by doctors – not their fault since, at the time, it was impossible to know that bacteria have a very clever way of becoming resistant to antibiotics.

We know now that a bacterium that is resistant to an antibiotic can pass this resistance on. This is how the immunity spreads. Many bacteria possess a little ring of DNA called a plasmid. These plasmids often have genes that code for antibiotic resistance, so exposure to the antibiotic would not harm the bacterium. What is clever is that the bacteria have the ability to make copies of these plasmids and pass them on to other bacteria (Fig 7.11). For this reason, in the fight against disease, bacteria are always one step ahead of us.

The discovery and understanding of the way in which plasmids work opened up a whole new world to scientists. The ability to manipulate these plasmids has led to advances in science and medicine unimaginable when Fleming made his observations.

Useful bacteria

Often, because of their association with disease, we remember only the bacteria that are harmful to us. The majority of bacteria are harmless and indeed many are essential to us, for example the bacteria that play vital roles in the nitrogen and carbon cycles.

6 Give a brief account of the role of bacteria in

a the nitrogen cycle
b the carbon cycle.

Biotechnology

Biotechnology could be compared to factory farming, using microorganisms instead of chickens. It means applying biology to the manufacturing industry. Although the term itself is relatively new, many of the processes have been with us for centuries. Mankind has exploited biotechnology in such activities as wine-making, brewing, bread-making and waste treatment. However, it has been the development of **genetic engineering** which is responsible for the present advances in biotechnology.

7 Make a list of any processes, other than those mentioned above, which you think involve biotechnology.

Food from bacteria

Pruteen – a high-protein food – is simply made from dried, crushed bacterial cells. It is used to make animal food. Bacteria are placed in a giant fermenter. They are supplied with a simple organic food, methanol, given a suitable temperature and a supply of essential mineral salts. The fermenter is kept sterile – no other microorganisms are allowed to contaminate it. The bacteria grow and multiply in the fermenter, converting methanol into bacterial protein.

8 Why do you think it is essential not to allow the fermenter to become contaminated by other bacteria?

Single-cell protein is a high-protein food made from a fungus such as yeast. It is fermented, with sugar as a food source, in a similar way to Pruteen (see Fig 7.12). This high-protein food might help solve the world food shortage, without having to breed more animals for meat.

Genetic engineering

Genetic engineering makes use of the fact that bacteria can be 'persuaded' to make a variety of products that they would not normally make.

There are many ways in which we can change the genetic make-up of an organism. We selectively breed plants and animals until we achieve the desired combination of genes. However, because of the complexity of these organisms, it is very difficult to manipulate directly the genes inside individual cells. If, for example, the genes in the pancreas were failing to produce insulin, it would be impossible to place a working gene in the nucleus of every one of these cells. Microorganisms, on the other hand, are unicellular and are much easier to manipulate. Genetic engineering, or recombinant DNA technology as it is also called, involves inserting a piece of DNA (a gene) into the bacterial DNA by using the following steps:

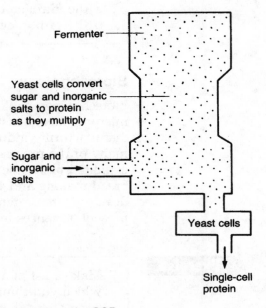

7.12 A fermenter used to make SCP.

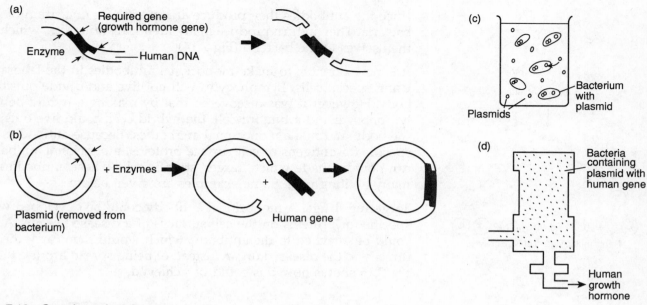

7.13   Genetic engineering.

1   The formation of the desired DNA fragment. This could be the gene for making human growth hormone, if that was the product to be synthesised. This can be done by using special enzymes to cut the gene out of human DNA (Fig 7.13(a)).

2   Plasmids are removed from bacteria and opened up using the same enzymes used to cut the human DNA. The human gene is then inserted into the plasmid, using another enzyme to stick it together (Fig 7.13(b)).

3   Bacteria and the altered plasmids are then placed together in a suitable medium. Some of the bacteria will take up the plasmids (Fig 7.13(c)).

4   The bacteria that have taken up a plasmid are then selected and placed in a fermenter. Given the right conditions, the bacteria will multiply and make human growth hormone in large quantities (Fig 7.13(d)).

Exercise

9 a   In the past, insulin for diabetics was obtained from animal sources. It was never wholly effective at treating the condition. Suggest a reason why.
  b   Human insulin is in plentiful supply today. Why?
  c   The ethics of genetic engineering worry many people. What kind of worries do you think people might have about the concept of manipulating genes?

Monoclonal antibodies

A similar technique has been developed to produce so-called **monoclonal antibodies**. When large numbers of identical cells are produced, they are called clones. When large numbers of one *type* of cell are produced, they are called monoclonal cells. For example, when the blood is invaded by microbes, this stimulates the rapid synthesis of B lymphocytes. Only one type of lymphocyte will multiply,

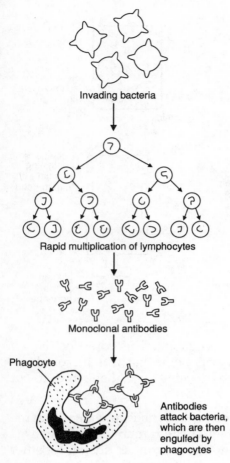

Invading bacteria

Rapid multiplication of lymphocytes

Monoclonal antibodies

Phagocyte

Antibodies attack bacteria, which are then engulfed by phagocytes

7.14   Lymphocytes making monoclonal antibodies in response to an infection.

since the antibodies they produce are specific for one strain of bacteria. They in turn produce many identical antibodies, which attach themselves to the bacteria (Fig 7.14).

It is now possible to make monoclonal antibodies in the laboratory, using special cells. Lymphocytes will not live and divide outside the body. However it was discovered that by making a hybrid between a lymphocyte and a tumour cell, the hybrid cells could live outside the body and make monoclonal antibodies. Because antibodies are specific for antigens or cell surface proteins, monoclonal antibodies can now be made which 'target' specific cells. These antibodies have many exciting uses. Some examples are given below.

Targeting drugs. A monoclonal antibody could be developed which is specific for proteins on the cell surfaces of a diseased organ. A drug could be attached to the antibody, which would then carry the drug directly to the diseased organ instead of being spread around the body. Fig 7.15 shows how this could be achieved.

Drug   Antibody   Diseased cell

7.15   Using monoclonal antibodies for targeting drugs.

Purifying proteins. A particular protein could be extracted from a mixture of proteins by developing a monoclonal antibody specific for that protein. The antibodies would bind to that protein only, allowing it to be removed from the mixture. In this way, the monoclonal antibodies act as a sort of biological 'magnet'. They can also be used in this way to detect the presence of certain proteins. If the protein is, for example, present in urine, the binding of the antibody to it could be used to bring about a colour change.

Exercise

10   How could monoclonal antibodies be used as a pregnancy test? **Hint:** during pregnancy specific proteins are found in the urine.

11   Cancer cells are often impossible to detect by X-ray. How could monoclonal antibodies help X-rays to track down a cancerous growth? **Hint:** metal atoms can be detected by X-rays.

Glossary

acetylcholine: A chemical that transfers nerve messages across a synapse.

acquired immunity: The immunity that we build up from birth as a result of encountering disease-causing organisms.

active transport: The movement of substances through cell membranes, often against their concentration gradient, using energy.

aerobic respiration: The release of energy from glucose, by combination with oxygen.

alleles: Alternative forms of a gene that are found at the same location on the chromosomes of a homologous pair. They are responsible for the same characteristic.

anaerobic respiration: The release of energy from glucose, without oxygen.

antibodies: Chemicals that are produced by special white blood cells (leucocytes) in response to antigens. They help to destroy invading bacteria.

antigen: The substance, usually protein, that is found on the surface of a bacterium. Different bacteria possess different antigens.

asexual reproduction: Reproduction without sex. It involves only one individual – no gametes or sex cells are involved and the offspring are genetically identical to the parent.

ATP: Adenosine triphosphate. A high-energy compound in which the energy release from glucose in respiration is stored. It is found in all living cells.

bacteria: Single-celled microorganisms that possess a cell wall but no nuclear membrane.

binary fission: The way in which bacteria for example continually reproduce by division.

biodegradable: A material that will be broken down and eventually decompose due to the action of microorganisms and other living cells.

biotechnology: The use of living organisms or biological processes for the manufacturing or service industries.

blood: Fluid found in vertebrates that transports nutrients, oxygen and waste products around the body.

breathing: A forced ventilation of the lungs to allow the gas exchange required for respiration.

cell division: Where a cell reproduces or makes a copy of itself.

cell wall: The structure which surrounds and gives shape to the cells of plants and bacteria.

chlorophyll: A green pigment found in plants which absorbs energy from sunlight to be used in photosynthesis.

chloroplast: Structure found in many plant cells which contains chlorophyll and where photosynthesis takes place.

compensation point: The light intensity at which photosynthesis and respiration occur at the same rate in a plant.

consumers: Organisms that consume other organisms for food.

continuous variation: The gradual transition between two extremes in a population, for example from very small to very tall.

cytoplasm: A jelly-like material surrounding the nucleus of a cell.

decomposers: Organisms (usually soil bacteria and fungi) that cause the decay of dead organisms and release mineral nutrients into the soil.

dialysis: The separation of a mixture of dissolved substances by diffusion through a semi-permeable membrane.

diffusion: The random movement of molecules, atoms and ions from a region of high concentration to a region of lower concentration.

diploid: Describes a cell or organism with a full set of chromosomes.

discontinuous variation: Type of variation where a character has two or more distinct forms with no 'in-betweens'.

DNA: Deoxyribonucleic acid. The material of inheritance – the genetic material of a cell.

dominant allele: The allele whose characteristic will be inherited by the individual in the presence of the recessive allele.

ecosystem: The living organisms and their environment in a certain area.

effector: An organ or cell that carries out an action in response to a stimulus.

endocrine system: A system of glands in an animal that produces chemicals (hormones) which are secreted into the bloodstream to control bodily functions.

enzyme: Biological catalyst; protein molecules made by living cells which speed up chemical reactions.

eukaryotes: Organisms whose cells possess true nuclei.

F_1 generation: The first generation from a particular cross.

F_2 generation: The second generation from a particular cross – the offspring of the F_1 generation.

flaccid: The state of a plant cell that has begun to plasmolyse and, because of losing water, become less firm.

food chain: A 'chain' that connects animals with their food and with the animals that eat them.

fossil fuel: Fuels made over long periods of time from the remains of once-living organisms, stored in the Earth's crust or at the surface as natural gas, oil, coal or peat.

gametes: Special types of cells that fuse during fertilisation. They contain half the number of chromosomes found in the other cells of the organism.

gametogenesis: Cell division that results in the production of haploid cells or gametes (meiosis).

gene: A unit of heredity. A segment of DNA on a chromosome which controls the development of a particular characteristic.

genotype: The genes of an individual.

gland: An organ that produces a specific hormone.

haploid: Describes a cell or organism with half the full number of chromosomes.

heterozygous: Describes an organism with different alleles of a particular gene, for example a person might have one allele that codes for blue eyes and one that codes for brown eyes.

homeostasis: The processes within the body that maintain a constant internal environment.

homozygous: Describes an organism with identical alleles of a particular gene, for example a person might have two alleles that code for blue eyes.

hormones: Chemical messengers that travel around the body of the organism. They are secreted from glands and affect specific 'target' organs or sites.

hydrostatic pressure: The pressure blood is under when entering a capillary. It is higher at the arterial end than at the venous end.

isotonic: Describes solutions that have the same osmotic potential because they contain the same salts dissolved to the same concentrations.

locus: The position of a gene on a chromosome.

meiosis: The type of cell division that gives rise to gametes or sex cells, which have half the full number of chromosomes. Sometimes called reduction division.

metabolic rate: The speed at which the chemical reactions of metabolism occur in an organism.

metabolism: The chemical processes taking place in a living organism.

mitochondria: Structures within cells that are the site of aerobic respiration.

mitosis: The type of cell division that produces an identical copy of a cell.

monohybrid cross: The production of offspring from two individuals that differ in only one characteristic.

mutation: A change in a gene or a chromosome that arises spontaneously.

nervous system: A network of cells that detect, and coordinate responses to, changes in the internal and external environments of an organism.

neurone: A nerve cell.

nucleus: The part of the cell, enclosed in a membrane, that contains the genetic material or DNA.

organ system: A group of organs that work together to perform a specific task, for example the lungs, kidneys, liver and skin make up the excretory system.

osmoregulation: Control of water content and salt concentration in the bodies of animals.

osmosis: The net movement of molecules through a semi-permeable membrane from a solution of higher water potential to one of lower water potential.

pathogen: An organism that causes disease.

pesticide: Chemical used to kill pests. Insecticides kill insect pests, fungicides kill fungal pests, herbicides kill weeds.

phagocyte: A cell that is capable of engulfing particles from its surroundings in order to destroy them.

phenotype: The physical appearance or characteristics of an organism.

photosynthesis: The process by which green plants make carbohydrates from carbon dioxide and water using energy from sunlight.

plasma: The fluid component of blood containing mostly water, solutes and antibodies.

plasmid: A tiny circle of DNA found in the cytoplasm of certain bacteria.

plasmolysis: Loss of water from a plant cell by osmosis.

producers: Organisms which produce food, i.e. green plants. They are the first organisms in a food chain.

prokaryotes: Primitive organisms whose cells do not have true nuclei.

pyramid of energy: A diagram which shows the amount of energy at each trophic level of the food chain.

receptor: A structure in an organism that detects a stimulus.

recessive allele: The allele whose presence can be masked or hidden by the presence of a dominant allele.

reflex: A fast, involuntary reaction to a stimulus involving at most two or three neurones.

respiration: The release of energy from carbohydrates which happens in every living cell.

response: A change or action that is brought about by a stimulus.

RNA: A long-chain molecule very similar to DNA which carries 'messages' from the DNA to the ribosomes.

root hair: Part of a root cell that projects into the soil, where it absorbs soil water and dissolved minerals.

semi-permeable membrane: A barrier which is porous, so that molecules which are small enough can pass through. Larger ones pass through slowly or not at all. The plasma membrane is semi-permeable.

sexual reproduction: Reproduction involving sex – the production of sex cells, or gametes, followed by fertilisation, giving rise to a unique individual.

solute potential: The ability of a solution to lose water by osmosis.

somatic cells: All the cells in an organism that have a full set of chromosomes.

spore: A small cell that can grow into a new organism.

stimulus: Any change in the internal or external environment of an organism that is large enough to be detected by a receptor.

stomata: Pores in the surface of a leaf that allow gas exchange to take place.

tissue fluid: Plasma minus the proteins and the cellular components of blood.

transpiration: The evaporation of water from the leaves of plants which provides a force to pull more water in the transpiration stream from the roots to the leaves.

trophic level: The level in a food chain at which an organism feeds.

turgid: The state of a plant cell that has taken in water by osmosis so that the cell contents have swollen and push firmly against the cell wall.

ultrafiltration: The process by which small molecules and ions are filtered out from larger molecules and cells in a liquid under hydrostatic pressure.

variation: The combination of new characteristics in an organism.

water potential: The ability of a cell to lose water by osmosis.

zygote: The cell produced immediately after fertilisation. When two gametes, for example a sperm and an egg, unite, a zygote is the diploid cell that results.

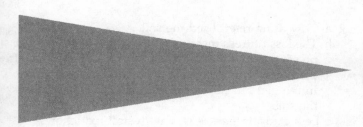

Answers

Chapter 1

1 Producers: thistle, seaweed, algae.
  Consumers: dog, blackbird, mushroom, coral.

2 a oxygen
  b photosynthesis (which produces the oxygen) only occurs in sunlight
  c water

3 a i Aerobic bacteria are around the chloroplast.
    ii Random small numbers of bacteria all around the *Spirogyra*.
  b Chloroplast. Aerobic bacteria are only around the chloroplast area which is in the light.
  c White, red and blue light can be used for photosynthesis.
  d Leaves appear green because they reflect and do not, therefore, absorb green light for photosynthesis.

4 a i no change
    ii rate roughly doubled
  b i no change
    ii rate roughly halved

5 a 6
  b i 2 to 6
    ii 9.5 to 10
  c 1 to 2

6 A CO_2 or sunlight; B CO_2; C low temperature;
  D sunlight

7 a i up
    ii down
  b 1 1
  c Deep inhalation followed by full expiration, then inspiration to full breathing.
  d 4 1
  e increasing exercise

8 a 5.5%
  b 15.5%
  c the atmosphere contains: O_2, approximately 21%; CO_2, approximately 0.04%

9 a i pH 8; ii 40 °C
  b Above 50 °C the enzymes become denatured.

10 a See Table A1.
  b decreases
  c Each allows more heat to escape from the elephant's body.
  d i Animal maintains a constant warm temperature irrespective of the environment.
    ii Animal can be more active on cold days and it can live in colder climates.

Table A1

Length of side of cube (cm)	Total surface area (cm²)	Volume (cm³)	Surface area/ volume ratio
1	6	1	6:1
2	24	8	3:1
3	54	27	2:1

  iii birds

11 a During vigorous exercise when there would be a shortage of oxygen in the muscles.
  b Poisonous lactic acid builds up (causing cramp).
  c i In the gut, where the tapeworm lives, there is little oxygen.
    ii There is no oxygen available during a deep dive.
    iii There is little oxygen available in the mud.

12 a i 10.00 a.m.; ii 7.00 a.m.
  b The bluebell grows in the shade of the oak at a lower light intensity. It must photosynthesise more efficiently at these lower light intensities.
  c The oak receives a higher light intensity and can photosynthesise more than the bluebell.
  d 10 cm³
  e 15 cm³ + 10 cm³ = 25 cm³ (the extra 10 cm³ oxygen allows for that used for respiration)

13 a There was no access to any 'extra' CO_2 in the sealed bottle. The plants' photosynthesis was always matched by their respiration so that no extra food product was available for growth.
  b After breaking the seal extra CO_2 was available to the plants so that they could now carry out more photosynthesis than respiration and so grow further.

Chapter 2

1 a Sunlight
  b Single-celled algae or cereal plant or leafy plant
  c Increase due to more food from crustaceans.
  d Decrease due to less food from snails.
  e Each organism uses food (energy) for its own respiration so that this food energy cannot be passed on to the next organism in the food chain.

2 a i A; ii B; iii D; iv C

3 a Different tissues containing different amounts of water would produce inconsistencies.

b (i)

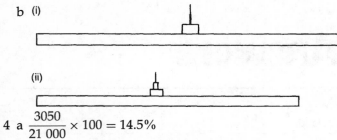

(ii)

4 a $\dfrac{3050}{21\,000} \times 100 = 14.5\%$

   b The bullock does not eat all the grass.
   c $3050 - (1020 + 1900) = 130 \text{ kJ/m}^2/\text{yr}$
   d Humans would eat vegetables directly. There is a greater loss of energy if bullocks are an intermediate in the food chain.
   e If we eat carnivores, then the food chain becomes longer (than if we eat herbivores) and this causes a greater loss of energy.
   f Native herbivores are likely to be better adapted to the local environment and therefore are likely to be more efficient in food production.

5 a 4 years   b 4 years   c 3 times
   d Increasing numbers of weasels are eating more field mice so the field mouse population decreases.

6 a The couch grass can live on loose sand that holds very little water. It can withstand salt spray and the occasional immersion under sea water.
   b Sand blown in the wind is deposited on the mobile dunes. Stormy winds bring in salt spray. There is salt but very little water in the sand due to fast drainage. There is little organic matter in the sand.
   c Planting with marram grass.
   d Marram grass thrives on being covered with sand on the mobile dunes. Once the sand has been stabilised by the marram other plants can invade the mature dunes.

7 a Soil bacteria and fungi.
   b Over periods of millions of years, shelled organisms died and fell to the bottom of the sea to form sediment layers. The calcium carbonate of their shells built up to form layers of limestone. Shelled molluscs were the last link.
   c Cement.
   d The burning of fuels increases CO_2 in the atmosphere. The loss of plants (e.g. deforestation) decreases global photosynthesis and so causes an increase in CO_2 levels. Increased numbers of animals (e.g. humans and cattle) increase global respiration and so cause an increase in CO_2 levels.

Chapter 3

1

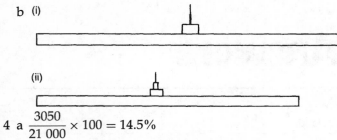

2 a Cow dung refertilised the soil.
   b The loss of trees means that tree roots are no longer present to bind the soil. The loss of trees has reduced the transpiration of water back into the atmosphere. Infrequent but heavier rainstorms now wash away the soil.
   c Less transpiration causes an overall reduction in rainfall since water is not being 'recycled' back into the atmosphere.
   d Cow dung is now used as a fuel, so there is less dung available to fertilise the soil.
   e Water cycle to show on-shore winds, evaporation of water from the sea to form clouds, condensation of the water in clouds to form rain over the land, transpiration of water from trees back into the clouds, the return of water in streams and rivers from the land to the sea.

3 a Nitrogen in nitrates combines with the carbon, hydrogen and oxygen of carbohydrates to produce protein.
   b Poor plant growth.
   c Phosphate is needed to produce ATP for the storage of energy in cells.

4 a In winter there will be less plant life and therefore less green algae in the lake.
   b Use a light intensity meter to measure light intensity at different depths.
   c Less green (chlorophyll) colour in the water will allow the light to penetrate further.
   d Look for an increase in nitrate/phosphate.

5 a Primary settlement tank: Organic solid sediments settle to the bottom of the tank leaving liquid above.
      Percolating filter beds: Aerobic bacteria on the stones break down organic substances in the liquid sewage.
   b To provide a large surface area for the aerobic bacteria to live on.
   c Sewage will drain from towns to a sewage works at a lower level. Treated water from the sewage works can be drained into the river.
   d Nitrates and phosphates can cause eutrophication (increased algal growth) in the river.
   e During warm weather there will be increased bacterial decomposition of the sewage leading to the release of more smell.

6 a bio = living. The biogas has been produced by the action of living microorganisms.
   b Residue material can be used as agricultural fertiliser.
   c It is prohibitively expensive to connect isolated farms and villages to a main sewage system.
   d Organic household waste such as paper, cardboard and kitchen waste.

7 a The suspended organic matter is gradually being used as a food source by the microorganisms in the river.
  b Some of the organic matter will be present in the sewage as suspended solids which are at a high level at the point of discharge. Organic matter will be respired by the river microorganisms causing the fall in the oxygen level that is shown on the graph.
  c Sludge worms and blood worms.
  d Stonefly and mayfly nymphs.

8 a cereal plant → pigeon → peregrine falcon
  b A falcon may eat several pigeons which have eaten lots of contaminated seeds, therefore the DDT accumulates along the food chain.
  c Some DDT may have drained off farmland into the sea to contaminate the food chains from single-celled algae to falcon.
  d Biodegradable insecticides break down and would not accumulate along the food chains.

9 calcium carbonate + nitric acid → calcium nitrate + water + carbon dioxide

$$CaCO_3 + 2\,HNO_3 \rightarrow Ca(NO_3)_2 + H_2O + CO_2$$

10 The snow over winter will act as a store of acid rain. When the snow melts a flush of acid water will enter the rivers killing the hatching trout and salmon.

11 Quarrying of large amounts of limestone will ruin the landscape of limestone areas in the countryside.

12 a Earth's crust store of limestone.
   b Calcium carbonate, from the shells of once-living organisms.
   c Living material store (increased respiration by more animals and reduced photosynthesis due to fewer plants).
     Earth's crust store (burning of fossil fuels).
   d 5.5 billion tonnes of carbon per year overall is being released into the atmosphere.
   e Increased rate of photosynthesis would cause a fall in CO_2 level in the atmosphere and may reduce/slow down global warming.

13 They are unreactive so will not readily react with other chemicals. Catalysts do not get used up in a reaction.

Chapter 4

1 Alveolus: CO_2 arrow out of blood into alveolus.
  Villus: glucose arrow into villus; amino acid arrow into villus; fat arrow into villus.

2 Large surface area; thin membrane(s); material that has been exchanged is removed from the area.

3 a +4.8; +4.5; +3.1; +1.0; +0.6; +0.1;
     −0.5; −0.8; −2.2; −3.3; −3.4

b

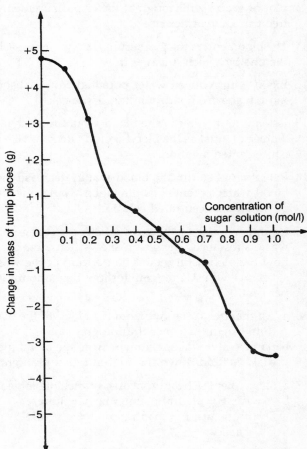

Change in mass of turnip pieces on soaking in different sugar solutions.

c approx. 0.52 mol/l
d Same answer as c. When the internal and external concentrations are equal, no osmosis takes place.
e i An increase in mass of approx. 0.9 g.
  ii A decrease in mass of approx. 2.7 g – by reading from the graph.

4 a Minerals dissolved in the soil water will cause it to have a lower water potential than pure water.
  b Water will move from the soil water to the lower water potential in the root hair cell.

5 a From cell B to cell A.
  b Equilibrium water potential = − 1.1 MPa

6 Sea water has a lower water potential than the root cells of garden plants and so, water will move from the roots into the sea water. The plants will wilt and die.

7 See glossary.

8 The end nearest the artery.

9 At the end nearest the vein. The hydrostatic pressure here is lowest and the opposing osmotic pressure is highest. This tends to draw fluid back into the capillaries.

10 Night-time plot of graph will be the same. Day-time plot of graph will trough at lower measurement for tree trunk circumference.

11 The blood enters the glomerulus in a wider vessel than the one by which it leaves it.

12 Blood with reduced water potential would absorb more water from surrounding tissue cells.

13 Retention of water in the blood would raise the blood water potential as the blood would effectively be a less concentrated solution.

14 Retention of salt in the blood would decrease the blood water potential as the blood would effectively be a more concentrated solution.

15 Counterflow gives the most efficient exchange of a waste substance. The graph shows that after parallel flow the blood still retains 50 per cent of the waste substance whilst after counterflow this amount is 30 per cent.

16 a All the urea can move from the blood to the dialysis fluid down the concentration gradient.
   b Some water will be removed from the blood into the dialysis fluid down the concentration gradient.

17 a air inside; trachea; bronchus; bronchiole; alveolus.
   b A = rib, B = pleural membrane, C = lung, D = bronchus, E = diaphragm, F = alveolus, G = trachea.

18 The volume of the lungs will increase.

Chapter 5

1 The person would be paralysed from the neck down, i.e. quadriplegic.

2 **Similarities:** Both have a cell body, axon, myelin sheath and dendrites.
Differences: The cell body of the sensory neurone is found along the length of the axon, in the motor neurone it is found at the start of the neurone (in the spinal cord). Sensory neurones carry impulses towards the spinal cord, motor neurones carry them away from the spinal cord.

3 a Stretch receptors in the muscle.
   b Stretch receptors → sensory neurone → relay neurone in the spinal cord → motor neurone → quadriceps muscle.
   c The lower part of the spinal cord, close to the top of the leg.
   d Approximately 110 cm (depending on the length of the leg).
   e If the distance was 110 cm, the speed would be 0.01 m/s.
   f The quadriceps contracts and its antagonist relaxes.
   g The stretch receptors detect pressure on the tendon. The purpose of the reflex is to prevent damage to the exposed tendon, when it is already stretched to its maximum limit.

4 a A = stimulus, B = receptor, C = sensory neurone, D = spinal cord, E = relay neurone, F = motor neurone, G = effector.
   b The finger would be withdrawn from the stimulus (drawing pin).
   c Involuntary.

5 a Eyelids shut (blink).
   b Tap the tendon just below the knee cap.
   c Thigh muscle.
   d Stand on a pin.
   e Muscles in leg.
   f Heat receptors in the skin.
   g Muscles in the hand and fingers.
   h Intercostal muscles in the rib cage and the muscles of the diaphragm.
   i Obstruction removed.

6 The next neurone would be stimulated continually.

7 These processes require energy in the form of ATP, produced by the mitochondria.

8 a In glass.
   b Follicle stimulating hormone.
   c The medium should have the same balance of ions as the body fluids. It should also contain the mother's serum for a source of protein and nutrients such as glucose.
   d The temperature and the pH should be kept constant.
   e Centrifugation of semen will cause the heavier part of the suspension, i.e. the sperm cells, to be forced to the bottom of the centrifugation tube. This lower portion will contain a very high concentration of sperm cells.

9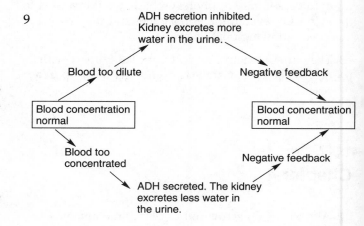

10 More water will be lost in the urine. This will cause the blood to become more concentrated, making the person more thirsty. The effects will be to cause dehydration, so drinking alcohol when you are thirsty is not a good idea!

1 a The cell walls must weaken so that they can elongate.
b Water moves into the cytoplasm of the elongating cells by osmosis.

12 Whatever is promoting growth is produced by the tip of the shoot.

13 The plastic is preventing the growth substance from diffusing down one side of the stem.

14 a The agar applied growth substance to one side of the tip only.
b The block that was in contact with the shoot tip for the longest length of time must have had more growth substance. This higher concentration caused a greater degree of bending.

15 a i 0.0001 p.p.m.; ii 10 p.p.m.
b Root cells are very sensitive to auxin. The concentration of 0.0001 p.p.m. stimulates growth, higher concentrations, far from stimulating growth, inhibit it.
c 1 p.p.m.

Chapter 6

1 a liver cell = diploid (2*n*)
b ovum = haploid (*n*)
c white blood cell = diploid (2*n*)
d sperm cell = haploid (*n*)
e muscle cell = diploid (2*n*)

2 a Fig 6.4; woman has 46 chromosomes.
Fig 6.5; woman has 47 chromosomes.
b She has an extra chromosome number 21, i.e. instead of inheriting a pair of chromosome 21s, she has inherited three.

3 a *C* = red, *c* = white. Choose the same letter (since we are referring to two alleles of the same gene) and not R for dominant red and w for recessive white.
b i *cc*; ii *CC*, *Cc*

4 See glossary.

5 *CC* × *cc*
Cc × *Cc*

	c	*c*
C	*Cc*	*Cc*
C	*Cc*	*Cc*

F₁ generation all brown.

6 a HH and Hh

	C	*c*
C	*CC*	*Cc*
c	*Cc*	*cc*

F₂ generation 3(brown):1(albino)

b The *h* allele is recessive. If it were heterozygous (*Hh*), the *h* gene would be masked by the dominant *H* allele. The only possible genotype therefore is *hh*.
c The genotype of the hairy-stemmed parent must have been *Hh*. *Hh* × *hh* gives the following combinations: *Hh*, *Hh*, *hh* and *hh*.

7 a AA × AB
b BB × AB

	A	B
A	AA	AB
A	AA	AB

50% A
50% AB
1:1 ratio.

c AB × AB

	A	B
B	AB	BB
B	AB	BB

50% A
50% AB
1:1 ratio.

	A	B
A	AA	AB
B	AB	BB

25% AA
50% AB
25% BB
1:2:1 ratio.

8 The man is responsible for passing on the Y chromosome to his offspring. Whichever sperm (a Y sperm or an X sperm) fertilises the egg will determine the sex of the baby. The chance of passing on a Y sperm is 50:50, but unfortunately Henry passed on an X sperm each time!

9 Because her mother would have to have been a carrier and her father a haemophiliac.

10 All her daughters are carriers (*Hh*); all her sons are normal (*H0*).

11 Beatrice: *Hh*, carrier; Henry: *H0*, normal.

12 **Mother** **Father**
  hh   *h0*
     or
  Hh   *h0*

13 *C* = normal vision, *c* = red–green colour blindness. Genotype of woman = *cc*, man = *C0*. Genotypes of offspring = *Cc* (female, carrier), *c0* (male, colour blind).

14 a discontinuous
b continuous
c continuous

d discontinuous
e continuous

15 Examples include pedigree dogs such as British bulldogs, which have been artificially selected and bred for their appearance. This has changed a lot, the very short, thick neck and broad chest being preferred by breeders. As a result, this breed often experiences breathing difficulties.

16 Asexual reproduction results in identical offspring, or clones. The ability to adapt to environmental change is dependent on new genetic combinations. The only way these occur in asexual organisms is by mutation.

Chapter 7

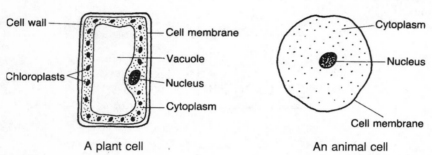

A plant cell   An animal cell

1 Find an agent that would bring about the destruction of the cell wall/membrane. Bleach works in this way.

2 When storing and preserving, make sure that the spores are unable to germinate, for example by freezing. When the food is defrosted it should be eaten immediately so as not to allow the number of bacteria to rise. Cooking may not destroy the spores.

3 Bacteria that can cause food spoilage. These are the same bacteria that recycle our nutrients. They are saprophytes and feed by secreting enzymes onto the surface of the food. The enzymes then digest the food, liquefying it in the process, so that the bacteria can absorb the nutrients. This is why the texture of food alters when it has 'gone off'. Another example is bacteria that cause disease – pathogens.

4 AIDS victims are unable to replace white blood cells. Every time they get an infection, the white blood cells that die in the destruction of the bacteria are not replaced. When the white blood cell count becomes dangerously low, the patient is defenceless and even a common cold can be a killer.

5 They must not be broken down by our digestive enzymes. This would render them useless.

6 a Nitrogen, an element essential to all animals and plants, is made available to them because of the work of bacteria in the soil. Bacteria help to break down proteins from dead animals and plants into ammonia, which is then turned into nitrates by the nitrifying bacteria in the soil. Plants take up the nitrates through their roots and make protein. This plant protein becomes a source of protein for animals.

b Some bacteria recycle nutrients by means of decay. For example, carbon dioxide from dead animals and plants is released into the atmosphere by saprophytes and fungi.

7 Butter manufacturing: A bacterium called *Streptococcus lactis* is added to give the butter a sour flavour.

Yoghurt making: Bacteria ferment the milk to produce the distinctive flavour.

Gasohol: Initially produced in Brazil, sugar cane is fermented to alcohol to produce a cheap, renewable fuel.

Biogas: Common as a source of methane gas in China and India. Bacteria of decomposition break down household waste in small containers called digesters.

Tissue cultures: In the right conditions, cells can be cloned in the laboratory and will develop into mature plants genetically identical to the plant from which the cells were taken (oil palm).

8 Other bacteria may ferment the methanol to produce undesirable products, some of which could be toxic.

9 a The correct structure of a protein such as the hormone insulin is essential if it is to function properly. Insulin from animal sources would be structurally different to human insulin and therefore less effective.

b Through recombinant DNA technology. The structure of the human insulin gene has been mapped and it has been synthesised and successfully incorporated into bacteria.

c There is a large number of issues that might cause concern. Possibilities include: manipulating bacteria for biological warfare; a source of human DNA is often from aborted foetuses; manipulation of human genes to give us features that are desirable in our offspring – 'playing God'.

10 Certain proteins are only present in the urine during pregnancy. A monoclonal antibody that is specific for these proteins would bind to them in the urine. The antibody–protein complex could bring about a colour change, giving a positive result if the urine changed colour.

11 A monoclonal antibody could be developed that is specific for proteins on the surface of cancer cells. A metal could then be attached to the antibodies, which would show up on an X-ray. The antibodies and their attached metals would accumulate wherever cancer cells are present.